# CAMBRIDGE LIBRARY COLLECTION

*Books of enduring scholarly value*

## Life Sciences

Until the nineteenth century, the various subjects now known as the life sciences were regarded either as arcane studies which had little impact on ordinary daily life, or as a genteel hobby for the leisured classes. The increasing academic rigour and systematisation brought to the study of botany, zoology and other disciplines, and their adoption in university curricula, are reflected in the books reissued in this series.

## The Rev. J.G. Wood

An Oxford-educated clergyman and prolific writer on natural history topics ranging from seashore wildlife to microscopy, John George Wood (1827–89) wrote and lectured for a receptive Victorian audience. His books were not rigorously scientific, but they made their subjects accessible to laypeople and were said to have inspired many future naturalists in their youth. His *Nature's Teachings* (1877) has also been reissued in this series. Theodore Wood (1862–1923) published this biography of his father in 1890. The account covers Wood's childhood and education, his clerical work and his desire to share his enthusiasm for the natural world with the public. His lecturing engagements, including a tour of America, and his home life are also discussed. An affectionate portrait of a significant figure in the history of popular science, this work sheds light on the intellectual interests of its subject and his readership.

Cambridge University Press has long been a pioneer in the reissuing of out-of-print titles from its own backlist, producing digital reprints of books that are still sought after by scholars and students but could not be reprinted economically using traditional technology. The Cambridge Library Collection extends this activity to a wider range of books which are still of importance to researchers and professionals, either for the source material they contain, or as landmarks in the history of their academic discipline.

Drawing from the world-renowned collections in the Cambridge University Library and other partner libraries, and guided by the advice of experts in each subject area, Cambridge University Press is using state-of-the-art scanning machines in its own Printing House to capture the content of each book selected for inclusion. The files are processed to give a consistently clear, crisp image, and the books finished to the high quality standard for which the Press is recognised around the world. The latest print-on-demand technology ensures that the books will remain available indefinitely, and that orders for single or multiple copies can quickly be supplied.

The Cambridge Library Collection brings back to life books of enduring scholarly value (including out-of-copyright works originally issued by other publishers) across a wide range of disciplines in the humanities and social sciences and in science and technology.

# PREFACE.

As it may fairly be claimed for my father that he was the first to popularise natural history, and to render it interesting, and even intelligible, to non-scientific minds, it has been thought advisable that some account of his life and labours should be prepared and published while his memory is yet fresh in the minds of those who have read his books or listened to his lectures. In the following pages, therefore, I have endeavoured to describe his three-fold work as clergyman, author, and lecturer, and at the same time to give a short account of his public and private life from his early boyhood to the closing days of his life.

Unfortunately for the labours of a biographer, the diaries which he left behind him—and which are by no means continuous—are extremely scanty, and often for many weeks together there is nothing but the barest entry of·work done and letters written, without amplification or details of any kind. By the aid of family information, however, I have, I think, been enabled to fill in the gaps; and I have only to ask that indulgence which all may crave who attempt the most difficult task of giving to the world the account of a father's life.

T. W.

BALDOCK, HERTS.
   *January*, 1890.

# CONTENTS.

## CHAPTER I.

### BIRTH AND EARLY LIFE.

## CHAPTER II.

### CLERICAL LIFE AND WORK.

## CHAPTER VI.

### LITERARY WORK (*continued*).

## CHAPTER VII.

### LITERARY WORK (*continued*).

## CHAPTER VIII.

### LITERARY WORK (*concluded*).

## CHAPTER IX.

### THE SKETCH-LECTURES.

## CHAPTER X.

### THE SKETCH-LECTURES (*continued*).

## CHAPTER XI.

### THE SKETCH-LECTURES (*continued*).

## CHAPTER XII.

### THE FIRST AMERICAN TOUR.

## CHAPTER XIII.

### THE FIRST AMERICAN TOUR (*continued*).

## CHAPTER XVII.

### HOME LIFE.

## CHAPTER XVIII.

### PETS.

## CHAPTER XIX.

### RECREATIONS.

# CONTENTS.

# THE REV. J. G. WOOD.

## CHAPTER I.

### BIRTH AND EARLY LIFE.

THE REVEREND JOHN GEORGE WOOD, clergyman,
author, and lecturer upon Natural History, father
of the writer of this memoir, was born on the
21st of July, 1827, in Howland Street, London.
His father, John Freeman Wood, a surgeon, and for
some years Chemical Lecturer at the Middlesex
Hospital, had three years previously married Miss
Juliana Lisetta Arntz, a young lady of German
parentage upon the father's side, who, having passed
the first fourteen years of her life at Dusseldorf, had
then completed her education, and finally settled in

B

England.  The first child of the marriage was still-
born ; and my father, who came next, was thus
practically the eldest of a family of fourteen.  Of these,
however, several died in infancy, and two more only
lived to early womanhood.

My father himself was a weak and sickly child
from his birth, and for several years, indeed, it was
never thought that he could possibly live to reach
maturity.  He suffered principally from violent attacks
of croup, which recurred at frequent intervals, and,
until he was eleven years of age, obliged him to be kept
under constant supervision at home.  Yet the child
managed to pick up a wonderful stock of knowledge in
spite of his delicate state of health, and was always
occupied in learning something, in some of the thousand
and one ways which presented themselves to his ever-
active mind.  Partly by instruction of the ordinary
character, and partly by a species of self-tuition peculiar
to himself, he learnt to read with wonderful rapidity
and facility, and at four years of age was thoroughly
familiar with the historical portions of the New
Testament, and was manifesting the first signs of the
extreme fondness for books which afterwards characterised the whole of his life.  He could not be kept
from them.  A book, merely *as* a book, had an intense
fascination for him, and he read with avidity almost
everything that came in his way, and not only read,
but remembered it.  Indeed, he always had a most
wonderful memory, except for dates and names, which
he could seldom recollect at all.  To the end of his

life he could cite *verbatim* long passages from books or poems which he had not read for many years, and apt quotations from all sorts of sources seemed to come to his lips without any effort of recollection whatever. And much of his success in literature was no doubt due to his marvellous power of extracting, as it were, at a single reading, the pith from the numberless books which he perused, and storing it up in some pigeon-hole of his mind until required for use.

Spelling, too, like reading, came naturally to him, for he possessed that curious side-shoot of artistic talent which enables one to *see* any required word in the mind's eye, without depending for the letters which compose it upon any mere effort of memory. Strangely enough, however, there were two words which always puzzled him, and to the end of his days he could never spell " cheque " without the addition of an unnecessary c, or "niece" without transposing the second and third letters. And, with regard to these two words, no amount of correction ever made the smallest difference.

Arithmetic, even in its simpler forms, was always beyond him. He did, no doubt, know that two and two make four, but I very much question whether he ever mastered the multiplication table. And certainly a piece of mere ordinary calculation was utterly outside his powers. Possibly this was in great measure due to the character of his early training. Mathematics, in the days of his youth, were little regarded, and sound classical knowledge was generally considered as the one

B 2

end and aim of education; and the arithmetical talent, if not cultivated in childhood, seldom attains to any degree of perfection afterwards. So that when my father had any sums to do, he always did them by deputy. Euclid, however, he liked, and often worked at it merely for the interest that he managed to extract from it. But that was the only branch of mathematical science of which he ever picked up more than the merest rudiments; and I have always had a shrewd suspicion that he kept no account of receipts and expenditure for the simple reason that he distrusted his own power of adding up his columns.

At four years of age the boy was taken to church for the first time; and there an amusing incident happened. He does not seem to have received any preliminary instruction in the Liturgy, and did not at all know what to expect when he entered the building. He behaved very well, however, and joined in the Lord's Prayer, which, of course, he knew by heart, with much reverence and devotion. By-and-by, however, the Lord's Prayer was repeated again, and this time he seemed a little bored, and took his part in it only under protest. But when the Litany drew near to its close, and the same Prayer was said for the *third* time, his patience came altogether to an end, and, rising from his knees, he sat down with an air of great determination, and a very audible remark to the effect that he "couldn't stand this no more!"

In 1830 it was deemed advisable, for more reasons than one, that the Chemical Lectureship at the Middlesex

Hospital should be given up, in order that the family might remove to Oxford. And there a house was taken in the High Street, which was subsequently vacated for another in Holywell Street, and that again in its turn for a third in Broad Street.

As the boy still continued very delicate, his father saw that the only chance for him was to keep him at home for the present, and to allow him to live as healthy and natural a life as possible. Outdoor exercise and amusements, therefore, were strongly encouraged, and the child learned to run and swim and climb with a facility which few boys of his own age could equal. In the water, more especially, he was always perfectly at home, and would tumble in backwards, or head foremost, and dive for eggs and three-penny pieces, and even play a sort of aquatic leap-frog, as readily as though the river were his natural home. Indeed, he spent much of his time on its banks or in its waters. There were trout to be tickled, crayfish to be caught, and creatures innumerable to be watched, and perhaps brought home for the aquarium. The spirit of emulation was rife, and every boy tried to do better than his fellows. And so each and all came to be as familiar with the water as with the dry land, never from the first having learned to consider it as an element to be dreaded.

The crayfish were caught in rather a primitive fashion. Paddling along in the water by the banks, the boys would carefully investigate every hole, until the long antennæ of the crayfish were felt projecting.

Then a sudden " grab " was made, the creature seized
behind the great claws, so as to deprive it of the power
of employing those formidable weapons upon the un-
protected hand, and forthwith *transferred to the cap,*
which in those days was a roomy article of attire,
capable of holding several crayfishes without danger
of overcrowding. The presence of half a dozen of
these creatures moving about upon the head, and
occasionally giving a sharp pull to the hair, does not
seem to have been regarded in the least, the great
beauty of the arrangement being, of course, that it
left the hands free, while there was little or no danger
of the captives escaping.

My father had many amusing stories to tell of
his early boyhood. One of an organised attempt to
excavate a subterranean passage from the garden to the
river-bank (half a mile away), which resulted in the
removal of huge quantities of earth, and the discovery of the scheme by the higher powers just in
time to prevent the probable burial alive of the
whole enthusiastic party. Another of a great plan
for the purchase of a donkey by means of the gradual
accumulation of halfpence; which plan seemed so
feasible, and so certain of fruition, that a big pair of
scissors were surreptitiously removed from the maternal
workbox, and the lawn diligently cropped, in order
that a store of hay might be laid up for the prospec-
tive animal's requirements. And a third of the queer
code of honour which forbade the plucking of apples
from the trees in the orchard (where windfalls were

recognised as common property), but did not militate against the employment of boys from outside to pelt the fruit with stones, by the bribe of a commission on the profits. " *Quod facit per alium, facit per se* " was a motto clearly unregarded by the youthful moralists. Very early in the boy's life the bent of his mind manifested itself; and he himself could never recollect the time when he was not constantly poking, and probing, and prying, here, there, and everywhere, in the endeavour to discover some of the manifold secrets of Nature, and to learn the ways and doings of the multitudinous living creatures that garden and river and woodland afforded.

In this he was much encouraged by his father, who, on Sunday afternoons, would lend a microscope and a pocket magnifying-glass to the children, and join eagerly with them in examining the numerous wonders which a few minutes' search in the garden would always turn up. Pets, of course, were numerous and varied. Bats, toads, lizards, snakes, blindworms, hedgehogs, newts, dormice, insects even of various kinds, all were kept in turn. And so the boy laid the foundation of that store of knowledge which afterwards served the man so well. He learned to love animals of all kinds, and to study with the deepest interest and minutest care every detail of their life-history. And at the same time he was unconsciously teaching himself how to observe, and learning the lessons, myriad and diverse, which Nature is always ready to impart to those who strive to search out her secrets.

Soon followed another step, and a most important
one, in the pursuit after knowledge, for at a very early
age the young naturalist found his way to the Ash-
molean Museum, and almost immediately succeeded in
getting upon unusually friendly terms with the kind-
hearted old curator, who sympathised most heartily with
the boy's keenness and wonderful thirst for information.
Any help that he could give was freely given, and soon
" Johnny Wood " was a constant visitor to the Museum,
and as constant an enquirer of the curator, who, so far
from being annoyed by his persistence, said that his
questions were so apt and sensible that it was a real
pleasure to answer them.   For several years these visits
were kept up, and even after school-days had begun the
boy's first visit at the beginning of every holiday season
was always to the Museum, in order that he might
discover all the new specimens, carefully examine them,
and find out whatever there was to be learnt concerning
them.

So passed the time until 1838, by which time eight
years of active, outdoor life, with unlimited exercise in
the way of running, swimming, climbing, and exploring
woodland, hill, and dell, had so strengthened the boy's
constitution that it was deemed that home study might
profitably be exchanged for the severer discipline of a
school.   He was therefore sent to Ashbourne Grammar
School, in Derbyshire, over which his uncle, the Rev.
G. E. Gepp, presided as head-master; and there he
remained for the next half-dozen years.

The school was conducted on old-fashioned principles,

all offences, great and small, being impartially visited with the rod, while the daily routine would now be considered as stern and rigorous to a degree. And the head-master, dreading to be accused of favouring his own nephew, was far more strict, and even merciless with him than with any of his fellow-pupils. Yet the six years which were spent there appear to have been by no means unhappy on the whole. There was plenty of time for outdoor exercise; the neighbouring country afforded every opportunity for the manifold forms of recreation in which the souls of boys delight; and, pleasantest of all, the natural history studies could be carried on almost as freely as at Oxford. Soon the boy collected about himself a band of kindred spirits, who used to scour the neighbourhood in search of specimens and trophies, and come home laden with spoil, both living and dead. Grass snakes more especially were in great request by way of pets. Almost every boy had quite a number of them, and would carry them about in his pockets, tie them round his wrists and neck, or cause them to run, or rather glide, races with those of his companions. A very favourite amusement, too, was to visit certain deserted stone quarries in the neighbourhood where standing water was always to be found, and there to make the snakes swim by the simple expedient of throwing them into the middle of a pool, and leaving them to find their way to land. Sometimes a snake would become obstinate, and lie sullenly at the bottom without attempting to swim; and then stones had to be thrown in such a manner as to fall close to it without injuring it. Sometimes even this plan would fail, and then there was nothing for it but to leave the snake master of the situation, and to go home without it. But generally there was little or no trouble of this kind, and snake-races could be conducted in the water almost as easily as upon dry land. The snakes very soon learned to recognise their masters, and to refrain from making use of the highly disagreeable odour with which Nature has gifted them as a means of protection against their foes. And, even when illicitly taken into school, they would lie quite quietly in the pocket without attempting to escape, or in any way giving notification of their presence.

I do not know that my father ever joined with any degree of enthusiasm in the ordinary out-door games of a schoolboy's life. He was something of a cricketer at one time, but, after his usual unlucky manner, contrived one day to catch his foot in a hole only a few inches deep, and, in the fall which resulted, to break his right leg rather badly and to dislocate his ankle. This involved confinement to bed for several weeks under peculiarly disagreeable circumstances, of which he gives a graphic account in his "Insects at Home," when speaking of that unpleasant creature, the common flea :—

When I was at school (he says), I had the misfortune to suffer a simultaneous dislocation and fracture of the ankle, and was conveyed to the infirmary, a large room at the top of the house. Now, this room had been without tenants ever since I remembered it, and I believe that for at least seven years no human being had entered the room, except to open the windows in the morning and shut them at night. The room was kept most scrupulously clean, and no one ever imagined that a flea was in it.

That the room was tenanted by these insects I found to my own proper cost. No sooner was the candle put out than a simultaneous attack was made on me in all directions. From every part of the room fleas came in battalions. There was a nurse in the room, who was one of those persons that are either impervious or objectionable to fleas, and she escaped them entirely, while they concentrated all their energies on me.

Now a damage such as I had suffered is not conducive to rest, even with all appliances. The limb swells, until the skin feels almost unable to resist the tension, and the burning heat is as if melted lead were being continually poured over the joint. Fever rages through the frame, and the first endeavour of the surgeon is to subdue it as much as possible. Under such circumstances, it may well be imagined that the ceaseless attacks of the flea armies were not calculated to produce quietude; and, indeed, had the occupier of the bed been in perfect health and strength, one such night would have sufficed to drive him into a fever. The only portion of the skin that escaped was that which was covered with the bandages, and even there the dreadful little insects had found out the junctions of the bandages, forced themselves under the edges, and driven their beaks into the skin, so that, when the bandage was removed in the morning, its course could be traced by the rows of flea-bites.

The insects had never enjoyed such a chance of a banquet in their lives, and naturally made the most of it.

After this highly unpleasant experience my father never seems to have taken any but a very occasional part in the game of cricket, although he retained his interest in it to the end of his life, and always studied the cricketing news in the daily newspapers with some degree of care. This accident, by the way, was the first of a long series. Seldom, I suppose, was there a man who injured himself more often, or with less permanent effect. He broke, at different times during his life, his right arm, his right leg, his collar-bone (twice), six ribs, almost all the bones of his right hand, and his nose. He cracked several other bones without actually fracturing them. He dislocated his ankle and several of his fingers. And yet the only lasting damage resulted from the injury to his right hand, which was of so serious and complicated a character that the only marvel is that he should ever have recovered the use of the member at all.

Remaining at school until he was seventeen years of age, he then returned to Oxford, and shortly after-wards matriculated at Merton College. In the follow-ing year he tried for and obtained the Jackson Scholarship; and partly by the aid of this, partly by taking pupils in his spare hours and during the vacations, he entirely supported himself throughout his university career.

In spite of his two-fold labours, however, he still contrived to keep up his natural history studies, both indoors and out. His rooms were full of cages, and nets, and boxes of all kinds. At one time he was studying the development of the tiger moth from the egg to the perfect insect, and had between five and six hundred of the "woolly bear" caterpillars simultaneously feeding in an enormous breeding-cage specially constructed for the purpose. This was so arranged that the stems of the food-plants passed through holes in the floor into a tank of water be-neath; so that while the caterpillars could not by any possibility suffer an untimely death by drowning, their food was kept fresh and wholesome. Yet, twice a day, so enormous was the appetite of the insects, an accommodating scout had to be despatched into the neighbouring lanes to bring in as big a bundle of dumb-nettles as he could carry. And this continued day after day, until all the caterpillars which remained were "full-fed," and ready to pass into the pupal or chrysalis state.

By this time, however, their numbers had been considerably diminished, for at regular intervals of a couple of days a certain number had been carefully bottled in spirits of wine; and so, when their growth was at an end, my father had a complete series in preservation, in all the stages from birth to maturity. These he subsequently dissected, and thus began his acquaintance with the very important and extensive subject of insect anatomy.

Other pets he had, too, at the same time: grass-snakes again, which had a way of escaping from their cage and lying up in all sorts of nooks and corners, to the great dismay of the "bed-maker" and the scout; bats, and various other creatures. About this time, also, he made a somewhat extensive collection of insects, principally consisting of butterflies, moths, and beetles, and worked the surrounding district very thoroughly, paying particular attention to Bagley Wood, which was always one of his favourite haunts. But yet he found time to join in many of the recreations of his fellow-students. He was very fond of

boating, and spent a good deal of his time on the
river. He was a most enthusiastic gymnast, and
became the most proficient member of the university,
as far as the bars, and ropes, and trapezes, and
vaulting-horses were concerned. He was fond, too,
of fencing and single-stick, and became no mean
proficient in the art of self-defence. Swimming, of
course, was kept up, as of old; and in the winter,
when the fields were flooded and the frost came, he
was on the ice at every available moment, practising
diligently at all the manifold varieties of figure-
skating, until he became an acknowledged expert in
every branch of the art.

He had many stories to tell of his college life;
a very strange one in particular, involving the disappearance of a poker, which, I believe, rests to this
very day deep down in the ground in the centre
of Merton "Quad." The adventure in question was
as follows :—

He was engaged in putting together the mechanism
of a small model steam-engine, and, finding himself
in difficulties, went off to ask counsel of a friend.
The friend gave the requested advice, and came out
of his door to wish his visitor farewell. No sooner
had the two crossed the threshold, however, than
the "oak" closed with a bang, and shut the occupant out of his rooms. Having left his latch-key
inside, there was nothing for it but to pick the
lock; and this, after twenty minutes' hard labour, the
two contrived to do. Upon entering the room, to

their utter surprise, they found it so full of tobacco-
smoke (which the occupant of the apartments cordially
detested) that it was impossible even to breathe or
to see until the window had been unfastened and
opened, and the fumes gradually expelled. Then, of
course, a search was instituted, and the puzzled
investigators found on the hearth a huge heap of
the coarsest and strongest tobacco, upon which was
laid a poker, which had evidently been lately heated
to redness. No sign was to be found of the
mysterious person who had placed it there. No one
was in the room; no one had passed out. The
windows were closed and fastened, of course on the
inside. The chimney was far too small to admit
of the ascent or descent of a human being, to say
nothing of the fact that a fire was burning in the
grate. And there was not even a water-pipe by
means of which an accomplished gymnast might have
climbed up the wall. The matter was a perfect
puzzle. For some time the two stood talking the
mystery over, discussing every possible expedient by
which the practical joker might have obtained admission to the rooms, and left them again before the
rightful occupant could return; and each in turn was
rejected as wholly impracticable. Thus half an hour
passed away, and again my father was accompanied
by his friend to the head of the staircase for a last
parting word.

No sooner had the two men passed the door
than the same programme was exactly repeated! The

oak slammed-to as before, fifteen or twenty minutes
were occupied in picking the lock, and when admission was gained the windows were closed and fastened,
and the room was once more full of smoke. When
the smoke had cleared away, the smouldering pile of
tobacco and the heated poker were found exactly as
before.  Not the smallest sign was to be found of
the perpetrator of the mysterious joke; not a trace
could be discovered of the manner in which he had
made his entry and exit.  The two men were completely at fault.

Then an idea struck the aggrieved owner of the
rooms. *Whose was the poker?*  It was a very ordinary
poker, with nothing whatever distinctive about it;
but it was not the poker which belonged to the room.
*That* was lying in the fender as usual, and had not
been meddled with.  Clearly the proper thing to do
under the circumstances was to send a scout round
the college on some pretext or other, in order to find
out whose fireplace was without its poker.  No sooner
said than done.  A scout was entrusted with the
commission, and visited every room; *but every room
had its poker.*

A council of war was then held, and it was agreed
that the owner of the mysterious implement should
never see his poker again.  So at midnight there set
out a solemn procession of two, one bearing the
poker, and the other the necessary tools for its inter-
ment: to wit, a crowbar, a wooden mallet, and a
heavy coal-hammer.  With the crowbar a deep hole

was made in the very centre of the college quad, and the poker placed upright therein. Then, with the mallet laid upon the top, in order to deaden the sound, it was driven deeply down by repeated blows of the hammer, until even the head was fully eighteen inches below the surface. Then the hole was carefully filled in, and the operators went off to bed. But no one ever applied for the poker, and nothing was ever heard of the clever joker who had laid his plans so carefully and so well.

On another occasion a siege was laid against my father's own rooms, which were quite at the top of the college, and approached only by a narrow and tortuous staircase. From an anonymous quarter, however, he received previous notice of the intended attack, and made all his preparations accordingly. First he laid in a large stock of grey peas, and a few long glass tubes, with a bore sufficiently large to carry them. Then he opened both the windows, so as to expose only half the surface of glass, protected that as well as possible, and finally procured a large " demi-john," filled it with water, and placed it at the head of the stairs ; and then he sat down to read.

About twelve o'clock sundry whisperings in the quad warned him that the attack was about to begin ; so he put out his lamp and waited. Next moment came a volley of stones, which were repelled by his fortifications ; and then he set to work with his pea-shooters. A little preliminary practice had made him a fairly expert marksman, and as soon as an assailant showed

C

hand or face, that hand or face was smartly struck with
a pea. The adversaries, too, laboured under the dis-
advantage that, although they could not see their
intended victim, whose room was in darkness, their
intended victim was perfectly well able to see them by
reason of the lights round the quad. So after a while
the enemy's forces were drawn off, a hurried consulta-
tion was held in a protected corner, and then a sudden
rush was made for the stairs. But on reaching the
last flight the expected victim was seen calmly waiting, with the demi-john of water at hand, ready to
deluge the first besieger who should be bold enough
to approach. The leader of the attacking party paused
and took in the situation; and then, with a laugh,
he remarked, " You fellows, I think we had better go
back." " I think you had," said my father; and the
enemy departed in confusion.

The three years of the ordinary college course came
to an end, examinations were safely passed, and in 1847
the future naturalist, still barely twenty years of age,
took his degree of Bachelor of Arts. Although not a
brilliant scholar, he had passed through his university
career with credit, and had imbibed a love for classical
learning which never left him to the end of his life.
Scarcely a day ever passed in which he did not read at
least a few pages of a Latin or Greek author. Horace
was always his favourite poet, and he was always picking up copies of his Odes at second-hand bookstalls,
at prices ranging from a penny upwards. Most of
these Odes he knew by heart, and would repeat them

to himself over and over again when lying awake at night. And he never lost an opportunity of advising others to read them, or of descanting with enthusiasm upon their manifold beauties.

Here is an old letter of his upon the subject written to one of my sisters. It was written from Boston, U.S.A., and is dated Christmas Day, 1883:—

As to the Horace, I have picked out some of the gems. They are tolerably easy, and it will be better for you to work at them instead of taking up the entire book.

Book I. Odes 1, 2, 3, 4, 5, 9, 20, 21, 22, 30, 37, 38.

Book II. Odes 13, 14, 16.

Book III. Odes 3, 9, 13, 26, 30.

They are songs *with* words—the delight of scholars, and the despair of imitators, the sublimest audacity concealed under a mask which is "childlike and bland." Now, I particularly want you to love Horace, as you have begun to love Shakespeare, and I hope you will love Chaucer and Spenser. As a rule, women get along well enough with Virgil, who was the Latin Tennyson; but Horace is too much for them. He took his measures chiefly from Alcæus and Sappho, and his Latin survives their Greek. Boil together Chaucer's "Romaunt of the Rose," Spenser's "Faery Queen," Shakespeare's Sonnets (with a few fiery flashes from "King John," "Henry V.," and the "Midsummer Night's Dream"), Keats' "Eve of St. Agnes," bits of Shelley's "Mab," Swinburne's classic odes, and Morris' "Earthly Paradise," and you may get a faint idea of the infinite variety, the unerring selections of unexpected epithets (not a "nice derangement of epitaphs"), the dainty choice of words, the burning patriotism, the gracious dignity of the scholarly gentleman, too proud to conceal his lowly origin; the self-respect of the poor man who could rebuke as well as praise Cæsar and Mæcenas, knowing that his life depended on the one and his living on the other—who, like "Hamlet," has no fault but that of being "made up of quotations."

My father's respect for the classical proprieties also showed itself occasionally in the strong protests which

c 2

are scattered throughout his writings against the ex-
reme looseness with which the terms used in scientific
phraseology are often framed. Take, for example, the
following, from his " Insects at Home " :—

I really do not like to translate such a word as *subapterus,* which
is a repulsive hybrid between Latin and Greek, and—with all
respect to the eminent entomologist who first manufactured it—
ought not to be accepted in its present form. What, for example,
should we think of such words as eightagon, twelvehedron, dreiangle,
petitscope, telesseer, insectology, etoilonomy, erdology, and the like ?
Yet there is not one of these words which is one whit more ridiculous
than subapterus. Should we be allowed to talk, much less write, of
a hemiglobe, an egg-positor, a chaudmeter, a baromeasurer, a virful
deed, or a megananimous sentiment ? But, if we are to retain the
one word, there can be no reason why we should not employ the
others.

Had the offending entomologist used the word *subalatus,* or
"partly winged," no one could have objected to it, as both words
are Latin. Apart from other reasons, it is a prettier-looking word
than *subapterus,* and much easier to say. But when he employs the
word *sub,* which is Latin, as a prefix to the Greek *pteron,* I do not
see that we should be called upon to excoriate our own ears and
those of future generations with such an atrocious compound.

I believe that brown sugar and oysters are considered incom-
patible, as is salt with strawberry cream. There is, perhaps, not one
in ten thousand who would not feel direfully aggrieved by having
any such mixtures forced on him as part of his daily diet. And
there is really no more reason for offending our eyes, ears, and mental
taste by *subapterus,* than our mere palates by the above-mentioned
mixtures.

During the whole of his university career my father
had studied with the special intention of taking Holy
Orders; but as he had matriculated so unusually early,
he was still barely twenty years old when he proceeded to

his degree as Bachelor of Arts, and was consequently obliged to wait at least three years longer before he could apply for Ordination. He therefore accepted a situation as tutor in a school of which the then rector of Little Hinton, in Wiltshire, was headmaster. Here he continued for two years, and was very successful in the work of tuition, while he imbued many of the lads with a taste for natural history. The half-holiday afternoons were commonly spent in long rambles over the downs, and in these two years he added considerably to his own zoological knowledge, and made many a note and observation which afterwards proved of the highest interest and importance.

In 1850 he left Hinton and returned to Oxford, in order to read for Ordination. Much of his time, however, was devoted to a private pupil, and as he was also working sedulously in the Anatomical Museum at Christ Church, under Sir Henry—then Dr.—Acland, the Regius Professor of Anatomy, two more years passed away before he was actually ordained. During these two years, to his great subsequent benefit, he went through a complete course of research in comparative anatomy, himself dissecting representatives of all the important families of the animal kingdom, and making numberless careful and valuable preparations, of which many remain in the museum to this day. Insect anatomy, in particular, received special attention at his hands, and he thus acquired an intimate knowledge of every part of an insect's structure, which afterwards stood him in more than good

stead. During these two years, in fact, was laid the
foundation of his future eminence as a naturalist. He
had previously, both as child and man, learned what
to observe in the way of outdoor zoology, and how
to observe it; he had gained a stock of personal
acquaintance with the ways and doings of birds and
beasts and reptiles and insects, in which at that time
he had few if any equals; and he had imbibed a true
love for the study of living nature, which drew him
to it purely for its own sake, and not by reason of
the future emolument which it might possibly bring
in. And now was gained the equally important knowledge of anatomy and classification. He learned to
understand on what principles animals are separated
into classes, and tribes, and orders, and families, and
for what reason those principles were chosen. He
learned to trace common characteristics in creatures
which to all outward seeming are separated far as
the poles asunder. And, above all, he came to understand the great and all-important law, that Structure
depends upon Habit, which afterwards formed the key-
note to so much of his writings. Without these two
years of careful study, he would never have been the
writer and naturalist that he afterwards became.
Probably he would never have taken to authorship
at all, unless, perhaps, as a writer for boys in boys'
periodicals; and certainly he could never have ventured
upon the large and important works which principally
brought his name into prominence. And the museum
itself also benefited considerably by his labours, of her

enriched its shelves with many a delicate and exquisite preparation, which perhaps brought out some detail of structure never before understood, while he also helped very largely in the systematic arrangement of its contents.

About this time, also, he was working very steadily with the microscope, in the use of which he became quite proficient, as evidenced by his " Common Objects of the Microscope," written some years later. For his insect dissections, of course, this instrument was absolutely necessary ; and during these and the few following years he prepared many hundreds of slides, and introduced several improvements of his own into the art of microscopic mounting.

In spite of all his zoological labours, however, and also of his literary work (for his first book—the smaller Natural History—was published in 1851), he kept up his reading, and in 1852, having obtained a title for the parish of St. Thomas the Martyr, in the outskirts of Oxford, he was ordained deacon by Bishop Samuel Wilberforce, who was at that time presiding over the diocese of Oxford.

# CHAPTER II.

## CLERICAL LIFE AND WORK.

Parish work—The Boatmen's Floating Chapel—Ordination as Priest—Resignation of Curacy—Chaplaincy to St. Bartholomew's Hospital—Marriage—Removal to Belvedere and resignation of Chaplaincy—Honorary Curacy at Erith—Old-fashioned services—An Explosion and its results—Organising a choir—"Aggrieved parishioners"—A burning in effigy—Presentment to the Archbishop—Cessation of opposition—Sole charge—Death of the Vicar of Erith—Subsequent clerical work—Style of preaching—Sermon notes—Maps and blackboards in the pulpit—"Flower Sermons"—Complaints of nervousness—Stammering cured—Last sermon—The Funeral Reform Association—Hatred of "mourning"—Work for the cause.

IMMEDIATELY after receiving ordination in 1852, my father threw himself heart and soul into his new work. His parish, which was situated in the poorest part of the city, was far from being an attractive one, but in a few months', time he had come to know every man, woman, and child residing within it, and was busily engaged in all the diverse labours which a parish of such a character entails. Besides serving as curate in this parish, too, he accepted the chaplaincy of the Boatmen's Floating Chapel, an institution in which he took the deepest interest, but which, of course, necessitated a good deal of additional labour. In consequence of all this heavy work (the services at the church were almost incessant, and all the curates were expected to attend them all), his application for priest's Orders had to be postponed until the end of the second year of his ministry; and then he received full Orders at the hands of the same bishop.

Shortly afterwards, however, owing to a variety of causes, he felt himself obliged to relinquish his curacy. The stipend attached to his office, in the first place, amounted to no more than sixty pounds a year, and out of this he was supposed to pay the interest upon the clothing club, and to make up the deficit in the salary of the schoolmistress, if the children's pence failed to amount to the stipulated sum. The laborious character of the work, and the necessity for constant visiting, prevented him altogether from adding to his income by the use of his pen; and so, in 1854, he retired temporarily from active clerical work, and went back to his literary labours.

For the next two years he took occasional duty only, often relieving a brother clergyman at one of the numerous Oxford churches. Early in 1856, however, he was advised to apply for the appointment of chaplain to St. Bartholomew's Hospital, which was then vacant, and with which was held also a readership at Christ Church, Newgate; and having done so, and interviewed several of the governors, he was shortly afterwards appointed to the post. On April 28th of that year he brought his long residence at Oxford to a close, and travelled up to London; and on Ascension Day, May 1st, he began his active work at the hospital.

There he remained for the next six years, during which he also carried on literary work with but little interruption. The duties at the hospital were not

arduous, save that they necessitated his residence within
five minutes' walk of the building, and that he was of
course always liable to be called upon to minister to the
spiritual needs of the dying at any hour of the day or
night. And so he contrived to find sufficient spare time
for his writing, without encroaching upon that which
was required for the duties of his sacred office.

Early in 1859, the monotony of his life was broken
by his marriage. In February, 1854, having come up
to town to attend a meeting of the Linnean Society, of
which he had just been elected a Fellow, he had met
Miss Jane Eleanor Ellis, fourth daughter of John Ellis,
Esq., of the Home Office, and a member of the Yorkshire branch of the family. An engagement soon afterwards followed, but was protracted for more than four
years. On February 15th, 1859, however, the wedding
took place; and my father always plumed himself
greatly on the fact that no single member of the
hospital staff knew anything at all about the matter
until it was all over. He simply left his rooms early
one morning, and returned a married man.

In 1861 he began to think seriously of giving up
his hospital appointment, and taking up his residence
permanently in the country; this for more reasons
than one. A child—a daughter—had been born a
year previously, and had died at the age of ten
months. A second child, born in 1861, was still-born.
My mother's health was in a very unsatisfactory state,
and he himself was far from well. Twice, indeed, he
had been visited with a species of blood-poisoning. On

the first occasion, serious mischief had been averted by
the timely use of the Turkish bath; but on the second,
an obstinate and painful gathering had formed on the
left hand, which did not show signs of healing until a
visit had been paid to Margate, supplemented by a
further, but shorter, trip to the New Forest. It was
evident enough that city life suited neither; and so, in
1862, just six years after coming to the hospital, he
sent in his resignation, and at midsummer migrated to
Belvedere, near Woolwich, where he remained for
rather more than fifteen years.

Soon after his arrival, he became acquainted with
the clergyman who was acting as *locum tenens* to the
vicar of the neighbouring parish of Erith, the Venerable
C. J. Smith; the vicar himself, who had formerly been
Archdeacon of Jamaica, being away from home for an
indefinite period. A kind of tacit agreement was
quickly entered into, in virtue of which my father
began to act as a kind of honorary curate, the parish
being a large one, and the duty somewhat too onerous
to be successfully undertaken by one individual. Not,
of course, that the ordinary week-day duties of a curate
fell to his lot: for those he had no time. But he
assisted in the Sunday and week-day services until the
return of the vicar in 1863, and then continued to do so
at the special request of the vicar himself.

The character of the services at this time was very
deplorable. The clerk's wife played the harmonium,
and the clerk did the singing. If a member of the
congregation ventured to join in the responses, or to

utter an *Amen* above a whisper, the remainder of that
body instantly turned and gazed with astonishment at
the offender. The chancel was squalid and dirty to the
last degree, the communicants at the monthly celebration of the Holy Communion averaged only five in
number, out of a population of some six thousand, and
the Church, to all appearance, was doomed to speedy
extinction as far as the parish of Erith was concerned.

So matters continued until 1864, in which year
occurred the memorable explosion at the Belvedere gun-
powder magazines, which stood upon the river-bank
about half-way between Belvedere and Erith. In addition to widespread and almost incalculable damage
spread over a wide area of country, this explosion so
wrecked the old parish church of Erith that, during the
necessary repairs, Divine Service had to be carried on in
the schoolroom. There music of a rather higher quality
was instituted, and, before the return to the church, my
father asked permission of the vicar to organise and
train a regular choir, and to provide properly practised
music at the Sunday services. The vicar gladly gave
his consent, and my father set to work to get the choir
together; no light task in such a parish, and with the
small amount of time at his command. Shortly after
the church was re-opened, however, a fully choral service was sung by a surpliced choir of fairly imposing
proportions. The harmonium was replaced by an
organ; the old slovenliness, formerly so painfully apparent both in building and service, was done away;
and bright hearty services began to attract regularly to

the church those who had previously deemed it unnecessary to attend Divine worship at all.

My father's share in the work of the church was now as follows. On Sunday morning and Sunday evening he said the prayers or sang the service; on Wednesday evenings he did the same; and occasionally —but very occasionally—he preached. After service on Wednesday evenings came the choir practice; and as a general rule, after service on Sunday evenings the choir adjourned to the Lady Chapel, and there sang a selection of anthems, less for the sake of the practice than as a sort of additional service of praise. After a time this custom came to be known and appreciated among the members of the congregation, many of whom would always stay for the singing after service. And the arrangement was most popular with the members of the choir themselves, who were thus enabled to indulge their taste for choral singing of a somewhat more advanced character, without the usual effect of destroying the thoroughly congregational character of the church services.

Yet the "innovations," as they were commonly styled, were not introduced without a great deal of opposition. Letters without number appeared in the solitary local newspaper of those days ; the clergy were freely accused of ritualism and Popery ; my father, as the originator of the surpliced choir, was even publicly burned in effigy. But the excitement gradually calmed down until the year 1867, when the malcontents were again aroused to indignation upon the occasion of a dedication festival.

This time the offending clergy were solemnly presented to the Archbishop, and were summoned to
Addington Palace, where the proceedings—which igno-
miniously collapsed in the end—were enlivened by the
laughter which followed the reading of one clause in
the indictment: " offertory collectors *in coloured bags.*"
After this little more was heard of the Erith " ritualism,"
and the constantly increasing congregation testified to
the favour with which the services were generally
regarded.

During the whole of the eleven years which elapsed
between his arrival at Belvedere and the death of Archdeacon Smith, my father rendered his services gratuitously, with two exceptions: the first for a period of
some six months, during the prolonged absence of the
vicar, who left him in sole charge; the second—in
1869-70 — for the space of a year, while the vicar,
owing to heavy family affliction, was travelling abroad.

On December 28th, 1873, Archdeacon Smith died,
after only two days' serious illness; and my father was
again left in sole charge until the appointment of his
successor. With this gentleman, unfortunately, he
found it quite impossible to work, and their views
indeed differed so radically and completely that he
ceased even to attend the parish church, and migrated
to a temporary district church which had lately been
erected in another part of the town. Here he occasionally officiated; but his regular clerical work had come
to an end for ever.

To the end of his life, however, he constantly exer-

cised his ministerial functions. While living at Norwood—from 1878 to 1885—he frequently assisted the
clergy of S. Philip's Church, Sydenham, and afterwards
was always ready to take Sunday duty to relieve a
brother cleric, and to give up his well-earned rest in
order that he might, in some degree, lighten the labours
of others. It was a common thing with him, while
absent upon his lecturing tours, to preach for a friend
upon the Sunday, wherever he might happen to be.
And as his sermons always cost him a vast amount
of care and anxiety, and, moreover, exhausted him very
considerably, the sacrifice upon his part, when consenting
to do so, was of no inconsiderable character.

His style of preaching was peculiarly his own, and
his sermons themselves were never like those of anybody
else. During the earlier part of his clerical life, he
always read from a manuscript; but afterwards, gaining
confidence by experience, he relinquished the practice
altogether, and trusted merely to the scantiest of notes.

I give here the outline of one of his later sermons,
exactly quoted from his notes; first, however, premising
that those notes are utterly incomprehensible to myself:

"Matt. v. 14, & vi. i.
" United . . . Church . . .
Worship . . . responsi-
bility. Not have to invent
. . . too much. Pharisees
did.
" Light united shine farther.
" Life.
" Quiet . . . Face of Moses.

. . . Effect of words and
deeds.
" Judge.
" Rebuke; not young old;
child parent ; *feel* awkward.
" Certainly, not *elementary*
duty.
" Keep away."

This is quite a fair specimen of the average style of his notes, which were generally written out in small handwriting upon half a half-sheet of note-paper, and upon which alone he depended as an aid to his memory while in the pulpit. How he even contrived to read them is a mystery, for he was very short-sighted, and could scarcely see at all without the aid of spectacles; how he managed to make anything of them when he *did* read them is a greater mystery still. But every single word in those brief jottings suggested some chain of ideas to his mind, which he had, of course, carefully thought out before, and which a "key-word," so to speak, would instantly bring before him again.

He was always very nervous in beginning a sermon; and generally the first few sentences, carefully prepared beforehand, were a little laboured and heavy. But then by degrees he would quite forget himself, and become wholly carried away by his subject; and the remainder of the sermon was always most instructive and striking. He well understood the use of those sudden, startling sayings which keep the attention of a congregation fixed, and cause them to hang on the lips of the preacher with a sort of breathless interest. I remember one occasion, for instance, in which he had been treating of the various phases of modern infidelity, especially as shown in the atheistical writings of a certain well-known platform orator; and his subject had led him to the question of the existence of the soul. "If," he said, "that man were to confront me, and to ask me whether or not I thought that I possessed a soul, I think that I should astonish him not a little by my answer. For if that question were put to me, I should say, No." Of course there was absolute silence in all parts of the church. Every eye was fixed upon the preacher who could give vent to such an appalling doctrine; every ear was eagerly waiting for the next words; the clergy in the chancel stalls were obviously most uncomfortable, and wondering whether or not such a statement ought to be permitted to pass unchallenged. Then he went on with his sentence. " Man has no soul. Man *has* a body ; man *is* a soul."

It was always a source of great regret to my father that the bonds of custom prevented him from using a black-board while preaching. He said that he could make himself understood so very much better if only he could illustrate his remarks with coloured chalks now and then as he proceeded, just as he did in his sketch lectures. He also longed at times to be able to hang up a map, and to have the pulpit formed rather after the fashion of a platform, so that he might walk up and down while delivering his sermon. Yet he was always one of the quietest of preachers, generally abstaining from even the slightest of gestures from beginning to end of his sermon, standing perfectly still, and seldom even raising his voice. He never ranted ; he never declaimed ; he never gave way to impassioned bursts of oratory. Just as in his lectures, he was plain and simple throughout ; the charm lay in the freshness of thought, the aptness of illustration, and the novelty

D

which somehow he contrived to impart to the most
familiar passages of Scripture.

He was, perhaps, especially happy in the " Flower "
sermons which have so much come into fashion of late
years; every member of the congregation being expected
to bring an offering of flowers, which, after being presented at the altar, is sent off for the adornment of some
hospital. His favourite text upon these occasions was
Isaiah xl. 6, 7, 8. I quote the following from an
account of one of these sermons preached at St. George's
Church, Ramsgate, on August 2nd, 1885 :—

The presentations having been completed, the Rev. J. G. Wood,
eminent as a naturalist, delivered a brief discourse appropriate to
the occasion. Selecting his text from Isaiah xl. 6, and the two
following verses, the preacher first of all reminded his hearers of
the beauty and perfume of flowers: God had filled the world with
beauty, showing them that beauty was a part of the Divine Nature,
so that they were bound in their little way to imitate God as well
as they could. And they had no excuse for not making the services
which they rendered to Him as beautiful as possible; for not giving
Him their very best, whatever that best might be. It was right
that they should fill their churches with beauty as far as they could,
and so give back to God something of those mercies which He had
showered upon them. Commenting on the words, "The grass
withereth, the flower fadeth, because the Spirit of the Lord bloweth
upon it," he said that they must remember that the Spirit which
blew death into the flower was the same Spirit as that which
breathed into it the breath of life. Then they must not forget that
there was diversity of flowers ; they saw all kinds of flowers, and
yet the same Spirit had breathed the life into them all. Again they
must remember that although the Spirit was the same, the means
which were employed in clothing that Spirit in bodily form were *not*
the same. This fact struck him very forcibly some weeks ago, when
travelling from the Isle of Thanet to Manchester ; he could not but

notice the extreme diversity of the flowers and plants in varying
atmospheres and soils. Successful gardeners took the trouble to
find out the soil which suited a plant best, and were careful to keep
it supplied with that soil. They must also remember that flowers
are not isolated. Why, it took all the laws of Nature put together
to make one flower. It took all the laws of chemistry to begin with
—such chemistry as the mind of man had scarcely dreamed of.
Flowers even that grew in the same ground had great diversity
among them. And what was the cause of that? It was the result
of the extraordinary chemical powers of Nature, brought into
beautiful form by a science which man had never been able to
discover, and never would discover. The science was the breath of
life which God had breathed into them. It took all the laws of
light to clothe the flowers in their beautiful colours ; and not only so,
*but every ray of light from the sun to the flower was a band that tied
that flower to the sun, and thus to the whole of the universe.* The
smallest daisy of the field was not isolated, but was a necessary unit
of the universe. Again, flowers had their work to do in the time
given them, and they were useless unless they developed into the
fruit of the future.

Having thus spoken of the flowers, the preacher endeavoured
to apply the truths thus drawn out. Children were the flowers of
humanity—they were in the early stage of growth—and people were
apt to look on them as they often looked on flowers, simply as pretty,
engaging toys ; and hence the number of what were called "spoiled
children." Let them look on children as the men and women of the
future, and remember that the responsibility rested with them as to
what kind of men and women they would ultimately become. The
flower faded because the Spirit of the Lord had blown upon it.
How often did they say to themselves : "O that the flowers would
not fade ! O that children would remain children !" And how foolish
they were ! Only a few weeks ago the Kentish orchards were full of
beautiful flowers, and they could but regret that the Spirit of the
Lord blew upon those flowers, and caused them to fall and die. But
they left their fruit behind them, and so performed the work for
which they were sent into the world. So with children. The Spirit
of the Lord blew upon them : the time was coming when they would
put away childish things, and they, as parents, could not but feel

D 2

regret, for they missed the merry patter of childish feet, and the hearty childish laugh. But although the child ceased to be a child—ceased to be a blossom—and even though the petals fell away, the plant did not die, and the individuality of the child did not cease. And why? Because "the word of our God shall stand for ever"—that word which equally breathed the spirit of life into the plant and into the child. See here the responsibility of parents. They were not to look on children as pretty playthings for the time being, but should try to instil into their hearts the word of God, so that, although the flower should fade, the word of the Lord should stand for ever. The responsibility was great, and they all had it to some extent. Even children had it to each other, and those who were older much more than children; and they would fail if they forgot it. The little child, of course, could feel none; but when children grew older they began to think—What shall I do in the life before me? Later on, when they got into active life, the idea in their minds, supposing them to be conscientious, was—What am I doing now? Am I doing the work which God has given me to do? When they passed into old age, then they asked themselves, What have I done?—and in all these cases there was a spirit of responsibility. They could not but feel, when young, that they should fail; when older, that they were failing; and in old age, that they had failed. And what of that? They were all human beings; who was there who had not failed? But suppose they learned this lesson—that God had breathed into them the breath of life—and acted up to the responsibility which that entailed; then they would know that their work would never fail and could never fail, because "the word of the Lord shall stand for ever."

This abstract is fairly complete and accurate; and yet it gives little true idea of the sermon. Its telling force depended so much upon the personal magnetism of the preacher; and no pen can transfer to paper the deep earnestness which made it what it was.

My father always used to complain that he was terribly nervous when preaching, but no one who did not know him very intimately indeed would have imagined for a moment that such was the case. So, too, in his lecturing, with regard to which he made a similar complaint. But I do not think that his nervousness ever lasted very long. His first few sentences were generally a little stiff and formal, and had obviously been carefully thought out and formed before the sermon began. But then, as he warmed to his subject, these traces of formality would altogether disappear; and I do not think that he was ever nervous after that.

His delivery was never very good. His voice was naturally rather throaty and husky, and at no time was it ever really strong. And yet he had the great faculty of making himself plainly heard, even in the most distant recesses of the largest building. Even when standing on the steps leading from the nave into the choir of Canterbury Cathedral, as he had to do when conducting the rehearsals of the great choral festivals, and issuing his orders to the choristers, who were but just entering from the cloisters, those orders were distinctly heard. Probably this was owing to the fact that he took such remarkable pains with the due enunciation of his words.

He had at one time stammered terribly, and although he had undergone a course of treatment, and had been almost completely cured, there were certain words which he could never utter without great and obvious difficulty; and he was even at times compelled to exchange these for others, from pure inability to pronounce them. Therefore, I think, he was the more careful with all his

words; and certainly even his most distant auditors
could always hear him quite easily and distinctly.

My father's last sermon was preached at Edenbridge,
Kent, on February 17th, 1889, when he selected 1 Cor.
ix. 9 as his text. Usually he placed the notes of his
sermon in his pocket Greek Testament after delivering
it; but as this sermon was preached after he had left
home for the last time, I am unable to find the brief
outline which he almost certainly wrote. And probably
it was accidentally destroyed with other private papers.

Connected with my father's clerical labours, although
not of them, was the work which he did in furtherance
of the objects of the "Funeral Reform Association."
He was himself a strong advocate of cremation, which,
as he used to say, only brings about in a couple of hours
the identical result which burial causes in a number of
years. Sooner or later the body must be dissolved into
gases, and he himself preferred that this should be done
by a process which involves no injury to the living, and
does away with some of the most repulsive circumstances
associated in the popular mind with death. Cremation,
however, not being advocated by the Association, which
aims principally at the simplification of funeral cere-
monies, and the speedy and true restoration of "earth
to earth, dust to dust," he set himself diligently to work
to further their aims; repeatedly speaking at their
public meetings, organising such a meeting at our own
house at St. Peter's, and losing no opportunity of en-
forcing their arguments both in sermon and in private

conversation. "Mourning," whether taking the form of black clothing, black-edged letter-paper, or the outward indications of woe which are usually so prominent at the modern funeral, was absolutely abhorrent to him. He could feel the loss of a friend deeply; but on religious, as well as upon other grounds, would never show his sorrow in the orthodox manner. And, long before he joined the Association at all, he wrote out careful directions for his own funeral, whenever it should come; expressing the very strongest desire that everything connected with it should be of the plainest possible description, that no lead coffin should on any account be used, and that no mourning should be worn for him by the members of his family.

This was a subject upon which he undoubtedly felt very strongly. It seemed so plainly evident to him that the more extravagant forms of mourning were utterly opposed to the spirit of the Christian religion—deeds giving the lie to words; and that the ordinary system of burial is merely a vain and reasonless attempt to delay that which is inevitable in the end. He recognised the sanitary side of the question, too, and urged the mischief often caused to the living by the unsatisfactory and illogical disposal of the dead. And so, as a question of religion, as well as one of plain common-sense, he did all that lay in his power to further the objects of the Association, and to enlist others in the cause.

Had his time permitted him, he would, I think, have taken up the question even more enthusiastically

than he did. There was nothing that he enjoyed more
than working up some subject upon which the public in
general needed enlightenment, obtaining all possible
information upon the matter, and then imparting the
results of his inquiry-to others. And he was so deeply
interested in funeral reform that he would have thrown
his whole heart into the work, and have done his very
best to bring that reform about.

# CHAPTER III.

## THE CANTERBURY FESTIVALS.

In the year 1867 the choir of Erith Parish Church, which had then been for three years under my father's management and tuition, was at its best. There was one point in which it was almost unique. All the adult voices were those of gentlemen, and the result was a refinement in the style of singing quite beyond the attainment of the ordinary village choir. Then the constant practices ; the minute attention paid to every detail of the service ; and, above all, the regular anthem-singing on Sunday evenings : all these had contributed to raise the choir to an unusually high pitch of excellence, and to render the church an attraction to visitors for miles around.

In the early summer of this year a dedication festival was held at the parish church of Northfleet, and both my father and his vicar were invited to take part. The preacher at the service was the Right Reverend Bishop H. L. Jenner, who had lately been consecrated to the See of Dunedin, but had not yet left England for his diocese. With him my father at once struck up a friendship, which afterwards ripened into intimacy; and Archdeacon Smith was so delighted with both the service and the sermon that he then and there resolved that a Dedication Festival should without delay be held in his own church at Erith, and that the Bishop, if possible, should again be the preacher. The Bishop, on being asked, at once consented, and in the following August the festival was duly held.

The music upon this occasion seems to have been unusually good, and the Bishop himself was very much surprised to learn that the very existence of the choir, as well as its excellence, was due to my father's labours. He was himself at that time the Precentor of the Canterbury Diocesan Choral Union, which held annual festivals—always on the Tuesday following Trinity Sunday—in the magnificent cathedral of the arch-diocese; and to this he made a passing reference when speaking at the luncheon which followed the Dedication Festival. There was one subject, he said, upon which he—the Bishop—would like to say a few words. He had ever regarded as a pet child the Canterbury Diocesan Choral Union, and in order to promote its success he had, with his colleague, always endeavoured to extend its workings throughout the county. He had even proposed to himself visiting every district, if possible, with a view to establishing branches. Other duties, however, had intervened which had prevented him from carrying out this idea in its completeness. He hoped that his remarks on this occasion would have the effect of inducing many of the parishioners of Erith to join this Choral Union, which could not fail to produce good results in perfecting the service of song in the house of the Lord. Then, after touching upon one or two incidental topics, he concluded his remarks by saying that there were very few country churches indeed in which would be found musical services conducted as they were at Erith.

Now, of course, the Bishop's rapidly approaching departure for his distant diocese involved the resignation of his post as Precentor of the Choral Union ; and he was at this time searching for some duly qualified man who might succeed him. The vacant post, it is true, was temporarily filled, but the holder was very anxious to resign it, and had, indeed, signified his intention of doing so after the following festival. And so the idea occurred to the Bishop that, if my father could produce such a service in such a parish, he would surely be the right man to occupy the vacant position. Quite unexpectedly, therefore, he one day paid us a visit, and, with the full concurrence of the cathedral authorities, asked him to take up the work which he himself had been obliged to relinquish. My father, after due consideration, consented. And so, in 1868, he found himself responsible at the cathedral for the greatest service of all the year, with the musical department resting upon

him, and him alone.  In other words,,he had to select
the music, arrange the order of service, communicate
with all the choirs in the diocese, and then travel from
one to the other for the space of eight or nine consecutive weeks, in order that each might receive the
benefit of his own personal instruction.  Then all the
necessary arrangements had to be made at the cathedral,
the order of service appointed, and the final rehearsals
held in the Chapter House ; so that, before the festival
could be held, nearly three months had of necessity to
be given up to the settlement of preliminaries, while an
amount of labour was involved which very few already
busy men would not gladly have avoided.

This, however, was by no means all, for, long before
the first practice could be held, the festival book had
to be compiled, and seen through the press.  And
this alone was an undertaking involving no little
time and trouble.  The music for two full services had
first of all to be selected, with the requirements of the
cathedral authorities, the choirs taking part, and the
service itself kept well in view.  Perhaps some special
chants or hymns had to be procured, and arrangements
made with composers.  Then the organist of the
cathedral had to be consulted, and perhaps also the
precentors of some of the principal choirs of the
diocese.  Then, when all this preliminary business
was over, the book had to undergo the process of
examination and revision by the Dean and Chapter.

Generally, that body took exception to some part
or parts of the book.  Then came a wordy warfare

through the medium of the post, usually carried on with a good deal of spirit, but resulting generally in concessions on both sides. After his first experience, I may here mention, my father used purposely to make an insertion or two which he himself had no desire whatever to uphold, and which he knew perfectly well would never be allowed to remain by the Chapter. That august body, however, usually remained content with the assertion of their power shown in striking out the objectionable passages, and allowed all that my father really wished for to remain unchallenged. And so all parties were satisfied.

Then, after the book was printed and published, arrangements had to be made with every choir which was to take part in the festival for a private practice by the precentor himself; necessitating a vast amount of correspondence, and the expenditure of much time and ingenuity in making the different fixtures work in with one another. A good deal of expense would also have been involved, but this was in great part obviated by the liberality of the South Eastern and London, Chatham, and Dover Railway Companies, who furnished the precentor with a free pass over the whole of their respective systems during the two months over which the preliminary practices extended.

But the mere necessary work attending the production of the festivals was sufficient to appal an already busy man, far more so one whose time was so greatly occupied as was that of my father. But he, nevertheless, went to work with his accustomed energy and enthusiasm, and at once set himself to raise the standard both of the music and of the actual service itself.

There was plenty of room for improvement in both. In 1868 my father attended the festival, and was much shocked to see the slovenly, and even irreverent, behaviour of those who, of all men, should have known better. Walking up the centre of the choir of the cathedral itself might be seen clergy, arrayed in full canonicals, carrying an ordinary tall hat in one hand, and with a gaily dressed lady on either arm. The alms at the festival service itself, instead of being presented at the altar, were deliberately and openly placed in a hat, and so carried off to the Chapter House. And all else was conducted on similar principles.

The combined choirs, again, numbered but some four hundred voices—a meagre show from a diocese comprehending more than as many parishes. And, finally, the festival service itself was of the most ordinary type; scarcely, in fact, superior to that which one may now hear upon every day in the week in almost every cathedral in England.

All this my father set himself to reform ; but, of course, he had to go to work carefully, and to do what he wished to do by slow degrees. Cathedral corporations are proverbially conservative and difficult to move; and argument, entreaty, sarcasm, invective, and bitter scorn were all freely employed without bringing about very much in the way of results. Perseverance and patience did their work, however, and after a time one or

two members of the Chapter suffered themselves to
be persuaded, and even took up the cudgels upon my
father's side ; and, although the warfare regularly broke
out year after year when the approaching festival came
up for consideration, most of the points for which
he contended were ultimately conceded.

In the first festival which he conducted—that of
1869 — he managed to secure a great accession of
reverence from all concerned ; and in that year, for the
first time, the alms were duly and properly offered upon
the altar by the present Bishop of Dover, who officiated.

His next step was to arrange for a processional hymn
—an undertaking in which he met with great opposition.  Hitherto the surpliced portion of the choir,
after robing in the Chapter House, had straggled hur-
riedly into the choir, mutely and untidily, and a great
and impressive effect had been allowed to slip.  Now
my father wished for a systematic procession, singing
some good and solid processional hymn.

His chief difficulty in arranging for this lay in the
attitude of the Dean (Dr. Alford), who, for a long time,
could not be brought to see that ordinary decorum
required an orderly procession, while such a procession
was hardly possible unless it were permitted to sing
upon the march.  Neither would he agree for a while
that the impressiveness of the effect was at all a thing
to be desired.  By dint of much perseverance, however,
my father carried his point ; and then incontinently
followed up his victory by suggesting that the Dean
himself should write a processional hymn for the occa-

sion, and compose the music also! The Dean, at first,
was a little overcome by the audacity of the proposal,
but finally consented; and shortly afterwards my father
received a very admirable hymn, with the Dean's com-
pliments. This, however, good as it was, was by no
means the kind of hymn which he wanted; and so he
wrote off again to the Dean, pointing out that the
hymn, while excellent in its way, was not at all adapted
to be sung upon the march. Would he kindly go into
his cathedral, walk slowly along the course which the
procession would take, and compose another hymn as he
did so?

The good old Dean was not in the least offended by
the unhesitating rejection of his work, and did as he was
bid; and the result was that grand hymn beginning
"Forward be our watchword," which, consisting of eight
twelve-line verses, has since been added to " Hymns Ancient and Modern," though set to different music. The
manuscript reached my father with a humorous little
note to the effect that the Dean had written the hymn
and put it into its hat and boots; and that my father
might add the coat and trousers for himself. On looking at the music, he found, accordingly, that only the
treble and bass had been supplied by the composer;
and, fearing to employ his own imperfect knowledge of
harmony in the attempt to supply the omission, he put
the matter into the hands of Mrs. J. Worthington Bliss
(Miss Lindsay), who kindly added what was necessary.

The effect of the hymn, when sung by the vast
body of choristers, was almost overwhelming. From

the time when the leaders of the procession emerged from the cloisters into the north aisle to that in which the last of the long stream ascended the steps of the choir, nearly half-an-hour elapsed. And throughout the whole of this time the glorious strains of Dean Alford's hymn were taken up again and again by fresh bodies of voices, each pair of choristers joining in the chorus as they reached a specified spot, and ceasing as they set foot on the last step of the ascent to the choir, and passed under the screen to their seats within. The effect of such a hymn, sung by such a body of voices in such a building as the grand old Cathedral of Canterbury, was utterly beyond the power of words to describe. Scarcely a member of the congregation but was visibly moved, and long before the last of the five " brigades," into which the choristers were divided, had entered the choir, it was felt that no such festival had ever before been held within the walls of the stately Norman building.

Of course this magnificent result was not obtained without an infinity of preliminary labour. And even the arrangements for the procession itself were exceedingly complicated. The original four hundred voices had now risen to more than a thousand, a very large proportion of which belonged to surpliced choristers. All these had to be so arranged that throughout the procession the due balance of the parts might be preserved; and at the same time some plan had to be devised, by means of which no member of the procession might at any moment be out of sight of the precentor's bâton.

E

The first of these difficulties was overcome by the division of the choir into the five "brigades" before referred to, each being constituted as a single choir, with the various parts in their proper balance. The second requirement was more difficult to fulfil, for there was no one spot in the entire nave where a conductor could be simultaneously seen by the whole of the procession. After much thought and experiment, therefore, it was determined that, during the processional hymn, my father should be assisted by three lieutenants, each armed with a bâton, whose duty it should be exactly to imitate his beat. He himself stood at the top of the choir-steps, while they were so posted that each could see him, and also that one of the four, at least, was visible from every part of the course which the procession was to traverse. And so the difficult question of time was settled.

As the last part of the last brigade set foot upon the steps leading to the choir the hymn was hushed, and they passed to their seats in silence. And then, as the precentor ascended his conductor's daïs near the lectern, the whole choir, surpliced and unsurpliced, broke out with one grand burst into the jubilant last verse, which was sung with full organ accompaniment.

In this year, for the first time, in view of the great strain upon the voices by the long service in the cathedral, and the preliminary practice in the Chapter House, the morning service was given up, and an afternoon service only held, beginning at four o'clock. Early in the day, however, there had been a choral celebration of the Holy Communion in St. Margaret's Church, the preparations necessary for the afternoon service preventing it from being held in the cathedral itself.

The preliminary practice in the Chapter House—the processional hymn itself being of course rehearsed *in situ* and on the march—occupied nearly two hours, and obtained the honour of a special notice in one of the leading musical newspapers. "There can be no doubt," writes the critic, "that, merely as a singing lesson, the practice in the Chapter House at Canterbury, under the auspices of Mr. Wood and Mr. Longhurst, was of untold advantage to the choirs. It was amusing to notice the astonishment of some of the rustics at finding out the real meaning of a 'rest,' and their evident satisfaction at the effect of the responses when sung with the proper pauses. It was clearly a new experience, a real revelation, to some of them. In like manner it is impossible to doubt that many young women were cured for good of their odious trick of 'slurring' one note into another by Mr. Wood's clever caricature of them, which made the Chapter House ring with laughter."

At one o'clock the assembled choristers were dismissed, to fortify the inner man against the fatigue of the afternoon; and at three o'clock the surpliced contingent again repaired to the Chapter House, in order to vest, while the remainder were ushered at once to their seats in the choir.

Towards the end of the festival service in this year a most striking incident occurred. The day had been a very dull and cloudy one, and, although no rain had fallen, the sun had never for a moment been visible.

E 2

The second of the two offertory hymns was that beginning
" Saviour, blessed Saviour," and at the commencement
of the sixth verse, just as the words, " Brighter still and
brighter glows the western sun," were being sung, the
sun broke out for a moment from behind the clouds,
pouring through the great stained windows upon the
mass of white-robed choristers, and flooding the choir
with light. It was only for a moment; but the effect,
coming just at that particular moment, and while those
particular words were being sung, was striking and impressive in the extreme; and no one who was present is
ever likely to forget it.

For the first time in the history of these festivals,
no sermon was preached upon this occasion, an omission
which provoked some amount of adverse criticism in the
press. But circumstances had so altered from those
of previous years that the change was rendered absolutely essential. When the choristers engaged only
numbered three or four hundred in all, and the congregation together with the singers could easily be accommodated in the choir, a short address was right and
proper enough; and Dean Alford, who was generally
the preacher, had a peculiar knack of saying the right
thing in just the right manner, while his clear penetrating voice was easily heard by all. But when the
number of choristers was nearly trebled, and every
corner of the choir was occupied by those who were
actually taking part in the service, some two thousand
people had to be accommodated with seats in the nave;
and how was it possible that they should hear a sermon

preached in the choir? The omission of the sermon,
in fact, was an obvious necessity, not a mere whim
upon the part of either the precentor or the cathedral
authorities, as some of the critics seemed to suppose.

Two years later my father managed to introduce
another improvement into the festival, in the shape of
brass instruments. These, however, were only employed
during the processional hymn, and consisted of four
cornets, the performers upon which led the procession,
and, on reaching the choir-steps, stood aside, still
playing as before, and allowed the long stream of singers
to pass between them. Then they too entered the
choir, laid aside their instruments, and joined in the
choral music as ordinary singers. The chief object of
the innovation was to support the voices, and to help in
maintaining them at the proper pitch. In former years
they had shown a tendency to become distressingly flat,
as was perhaps only natural in a hymn of such length;
and once, after beginning in the key of G, the processional was finished in that of F. This tendency the use
of the cornets entirely obviated; and the hymn went
better than ever before. In this year the choir numbered no less than twelve hundred voices, and the
proportion of surplices had considerably increased.

In 1875 my father conducted the festival for the
last time. He was beginning to find that he could no
longer manage to give up the two months necessary for
the preliminary practice, or afford the expense which, in
spite of the liberality of the two railway companies,
naturally attended the incessant travelling from place to

place; and he therefore reluctantly sent in his resignation, which was as reluctantly accepted.

He had done much in his seven years of office. He had secured at least outward reverence before, during, and after the service. He had raised the general standard of the music. He had greatly improved the performance of that music. He had introduced the processional hymn, and the brass instruments. He had brought up the numbers of the choir from four hundred to three times that number. And, incidentally, he had raised the tone of choral music throughout the diocese, and indirectly facilitated its introduction in parishes where it had never been known before. In relinquishing his bâton, therefore, he could feel that he had done his work; but I am sure that he deeply regretted the necessity of doing so, and that he would have been only too glad to continue that work if such a course had been at all possible.

# CHAPTER IV.

THE first idea of taking up literary work as at least a supplementary profession appears to have occurred to my father some time during the year 1850. At that time, having given up his tutorship at Hinton, he was residing in Oxford, and occupying himself partly in the tuition of a private pupil—with whom he afterwards paid two short visits of a few weeks each to France— partly in studying comparative anatomy under Doctor— now Sir Henry—Acland, the Regius Professor, and partly in reading for Holy Orders. Probably he felt that it would be well, if possible, to obtain some pecuniary profit from the work in which he was so much absorbed; and his rapidly increasing familiarity with the wonders of the animal kingdom gave him good ground to suppose that he could produce a book which would at least be accurate as far as the subject-matter

was concerned, and which might very possibly help to
instruct the public upon a branch of science of which
mankind in general then knew very little.

For in those days the book of Nature was practically
a sealed volume to all but the few who were able and
willing to undergo a long apprenticeship before they
could become acquainted with its marvels and its
mysteries. It had been made a hard, dry science,
teeming with technicalities and incomprehensible
phraseology, and sesquipedalian and often unmeaning
nomenclature. Classification was regarded far more
highly than the study of habits and life-histories, and
animals were looked upon, in fact, rather as cleverly
constructed machines than as living beings made from
the same clay as man himself. And consequently
Natural History had come to be associated in the
popular mind with all that was uninteresting and
repellent, and the wonder-world of Nature, only needing the easily applied key of Interest to open it, was as
yet almost wholly unknown.

So my father set himself to write a small Natural
History for the general reader, in which technicalities
and scientific phraseology should be either set aside
altogether, or at least, when necessity compelled their
adoption, be carefully and simply explained. From
this principle, in fact, he never swerved throughout the
whole of his literary career. Thoroughly familiar himself with the rules of classification, and perfectly at
home in the tongue not "understanded of the people,"
which was then almost invariably adopted by writers

upon Natural History, he could yet thoroughly appreciate
the manifold difficulties which they presented to others,
and especially to such as were but just entering upon
the first rudiments of the science. And so he resolved,
as far as his own writings were concerned, to use only
simple and plainly intelligible language, which, with no
parade of learning, should yet convey accurate knowledge upon the subject of which it treated. And I
do not think that in any of his books or magazine
articles there is one single sentence which could not
easily be understood.

The book appeared in 1851, under the auspices of
Messrs. Routledge & Co., and met with a sale which,
if not phenomenal in its character, amply justified both
author and publisher in undertaking further ventures.
The first step towards popularising Natural History had
been taken, and the public had responded, if not with
ardour, at any rate with warmth. And my father
began to feel that a literary career was before him, and
a definite line of work laid down.

For some time after the production of his first
volume, however, he was prevented by the force of
circumstances from following up his success. His pupil
naturally took up much of his time; his anatomical
studies, which of course he could not regulate to suit
his own individual desires, occupied still more; and to
the preparation for his Ordination, which was now
drawing near, he was obliged to devote several hours
of daily labour. And all that he could do for a while
was to collect material, and to write a few lines when-

ever he could contrive to find a little leisure-time. Yet
he managed to translate from the French Alphonse
Karr's charming "Tour Round my Garden," and to
bring it out with divers editorial notes. This was in
1852, in June of which year came his Ordination;
and then for two years he was busier than ever. The
work of the parish took up almost the whole of his
time; every hour of every day had its own special
duties assigned to it. And literature had, of course, to
go to the wall.

In the following year, however, appeared the first
volume of "Anecdotes of Animal Life," which had
been written mainly before his Ordination, and completed in odds and ends of spare time afterwards. The
title of the work explains itself, as far as its general
idea is concerned; but, so far from being in any way
comprehensive in its scope, it was limited to some eight
or nine animals only, which were treated at considerable length, in anecdotal manner, and discussed most
thoroughly from different points of view. In 1856
appeared the second volume of the same work, in which
the same system was adopted with another group of
animals, both volumes meeting with a very fair measure
of success. The two have since been published together
under the less happy title of "Animal Traits and
Characteristics."

His next literary work was the editing of "Every
Boy's Book," for Messrs. Routledge & Co., a task for
which his own skill in almost all outdoor and indoor
sports eminently fitted him. And then—in 1854—he

began to find, as shown in the preceding chapter, that the arduous and poorly paid parish work must be given up for a time, and literature be regarded awhile as the crutch instead of as the staff.

The perennial " Bird Question " was now occupying his thoughts a good deal, and though he seldom, during his career as an author, approached Natural History from its economic side, he began industriously to collect information respecting the influence of birds on agriculture and horticulture, by way of supplement to his own experiences of very nearly twenty years. As a result of this study, he found himself able to champion the cause of the birds, and, towards the end of 1856, " My Feathered Friends " embodied the result of his investigations, and pointed out the extreme value of the smaller birds alike to gardener and farmer. Blackbirds and thrushes, it was shown, although they eat a certain amount of garden fruit, amply atone for their occasional mischief by the vast amount of snails and noxious insects which they destroy. Some of the finches are fond of corn ; but then, on the other hand, they feed themselves partially, and their young entirely, with some of the most troublesome and mischievous of all the farmer's foes. And so, though undoubtedly injurious at one season of the year, they are as undoubtedly beneficial at another. The rook steals walnuts and potatoes, and also visits the corn-stacks at times ; but then the benefit which the same bird confers upon the farmer by the wholesale slaughter of wire-worms and other root-feeding grubs is simply

incalculable. The kestrels and the owls, in spite of the
accusations so freely brought against them by game-
keepers and owners of poultry, are altogether invaluable
benefactors, and alone prevent the produce of our
fields from being entirely destroyed by mice. And
even the much-vilified sparrow is not altogether so
black as he is painted, but undoubtedly possesses more
than the one redeeming virtue to qualify his thousand
crimes. Such was the teaching of "My Feathered
Friends."

In the following year—1857—Messrs. Routledge &
Co., who had conceived the idea of publishing a series
of shilling Handbooks on Natural History and kindred
topics, requested my father to undertake one at least of
the volumes ; and he, therefore, set busily to work upon
"Common Objects of the Sea-shore." The book was
not a large one, and the actual writing was a matter of
only a few weeks; but, as he did not care to describe
any animal with which he was not thoroughly familiar,
the preliminary investigations occupied some little time,
and the small sum which he received for the copyright
was certainly thoroughly earned.

The book appeared towards the end of 1857, and
met with an immediate and marked success, the
publishers being scarcely able to keep pace with the
demand. It was quite a new thing for those who make
holiday at the seaside to be able to learn something
about the various creatures which they were daily
finding in the rock-pools, or lying dead upon the
shore ; and the little handbook opened out quite a new

world, while the popular style in which it was written
rendered it easily intelligible to all.

In connection with this book my father met with a
rather amusing incident. Soon after its publication, he
was hard at work among the rock-pools at Margate, a
mallet and a chisel in his hand, his oldest coat on, and
his trousers tucked up to his knees. Just as he was
moving from one pool to another, a small company of
fashionably dressed young ladies approached, deeply
intent upon a copy of his own "Common Objects."
Just as they passed they looked up, saw the enthusiastic naturalist in his working attire, shrugged
their shoulders, elevated their noses, and murmured,
"How very disgusting!" And then they returned to
their book.

The success of "Common Objects of the Sea-shore"
was followed by still more striking results in the case of
"Common Objects of the Country," which appeared in
1858. The book took the public completely by storm.
A first edition of one hundred thousand copies was
prepared, and at the end of a single week not a copy
was to be procured! Edition followed edition, and still
the printers and binders could scarcely work with
sufficient rapidity to meet the orders which still came
pouring in. After a time, of course, the demand
slackened; but from that day to this it has never
ceased, and "Common Objects of the Country" is still
a book which commands a yearly sale.

Most unfortunately, however, my father, when
making arrangements for the production of these two

books, accepted the same terms which were offered to
the writers of other books of the same series, and
disposed of the copyrights for merely a small sum.  He
could not, of course, foresee the astonishing success with
which the books would sell, and, looking rather to the
length of time occupied by the actual preparation of the
MS.—of course, only a very few weeks—than to the
return which those books would bring in to the
publishers, took what was offered him, and parted with
all further interest in the publication.  Had he retained
the copyrights, there can be no doubt that he would
have cleared a large sum of money ; as it was, the actual
remuneration which he received for each of the two
handbooks amounted to only thirty pounds !

After the first of the two little books was published,
a great number of letters reached him from readers,
most of them asking for further information upon
certain points, and some of a very amusing character.
Perhaps the funniest was one dated from Cincinnati,
U.S.A.  The writer had read the Rev. J. G. Wood's
interesting book with much pleasure ; but, living so far
from the sea as he did, many animals described therein
were absolutely unknown to him.  And, in particular,
he had a great desire to examine a jelly-fish.  *Might* he
ask the Rev. J. G. Wood to forward him one by return
of post ?

About this time, appeared "The Playground," in
which my father—I believe for the first and only time
as far as book-work was concerned—adopted the *nom
de plume* of "George Forest."  The little volume in

question, too, represents his only venture in the direction of fiction, the book being a small tale of school life, so constructed as to give, in narrative form, much useful advice upon outdoor and indoor games, and athletic sports of various descriptions. In one of the characters—Edward Benson, eldest son of the head-master—he depicts himself as he was when a young man; small and slight, and apparently weak and unhealthy, but with great power of endurance, and no little development of muscle. The book is so arranged as to include exactly a year of school-life ; so that the sports and recreations adapted to the different seasons are all described in due succession. It has now, I believe, for many years been out of print.

The phenomenal success of the "Common Objects of the Country" led to arrangements for the production of a very much larger and more important work—the second Natural History. The preparations for this were made upon an unusually lavish scale. All the illustrations were to be drawn specially for the work, and only the best artists were to be employed. Type, paper, and all the other accessories were to be of the best description, and no expense was to be spared either in production or in advertising. Finally, the work was first to make its appearance in monthly parts (of which there were to be forty-eight in all), and, after the whole was completed, it was to be re-issued in the form of three bulky volumes, of large octavo size.

Of course the labour connected with the publication of this large and important work was very severe. Each

month my father was responsible for forty-eight pages
of letter-press—due deduction being made for illustrations; and each month many hours had to be given up
to personal interviews with the artists, correction of
blocks and printer's proofs, and all the manifold details
connected with the production of any work upon a
tolerably large scale. Then every available source of
information had to be sought out; all the leading
authorities examined; new material obtained from those
who had any personal knowledge of the rarer animals
described; and almost daily visits paid to the Zoological Gardens in the Regent's Park, and the leading
London museums. And all this in addition to the
labour involved by the actual writing.

Into this book my father put perhaps his very best
work. All who know the three stout volumes will be
able to appreciate the careful labour bestowed on the
description of every individual animal; but over and
above this there is much of a higher quality, much in
which a deeper note is struck, and in which some of the
many problems as yet unsolved by man are brought
forward, treated with reverent care, and finally put by
with an evident sense of regret. Perhaps I may be
permitted to quote the following by way of example:—

The attribute which we call Destruction ought to be termed
Conservation and Progression, for without its beneficent influence all
things would be limited in their number and manifestation as soon
as they came into existence, and there would be no improvement in
physical, moral, or spiritual natures. In such sad case, it would be
possible to find a centre and circumference to creation, whereas it is
truly as unlimited as the very being of its Creator.

Suppose, for example, that the huge saurians of the geological eras had been permitted to retain their place upon the earth, and that the land and water were overrun with megatheria, iguanodons, and other creatures of like nature. Suppose, to take our own island as a limited example, that the land were peopled with the naked and painted savages of its ancient times, unchanged in numbers, in habits, and in customs. It is evident that in either case the country would be unable to retain the higher animals and the loftier humanity of the present day, and that in order to escape absolute stagnation it is a necessity that old things should pass away, and that the new should take their place. How limited would the human race be were it not subject to physical death ! But a very few years and the earth would be over-peopled, setting aside the question of bodily nourishment, which requires the destruction of other beings, either animal or vegetable. The same rule holds good with regard to moral as well as physical improvement, for it is necessary that all mental progress should be caused by a continual destruction, a death of erroneous ideas, before the corresponding truths can obtain entrance into the mind.

Apply the same principle to the entire creation, and it will become evident that the destructive attribute is essentially the preserver and the improver. Death, so-called, is the best guardian of the human race, and its preserver from the most terrible selfishness and the direst immorality. If men were unable to form any conception of a future state, and were forced to continue in the present phase of existence to all eternity, they would naturally turn their endeavours to collecting as much as possible of the things which afford sensual pleasure, and each would lead an individual and selfish life, with no future for which to hope, and no aim at which to aspire.

The popular error respecting the destructive principle is that it is supposed to be identical with annihilation, than which notion nothing can be more false in itself, or more libellous to the Supreme Creator of all things. Death is to every man a terror, an abasement, or an exaltation, as the case may be ; but, in truth, to those who are capable of grasping this most beautiful subject, destruction is shown as transmutation, and death becomes birth. Nothing that is once brought into existence can ever be annihilated, for the simple reason that it is an emanation of the Deity, who is life itself, essential,

F

eternal, and universal. The form is constantly liable to mutation, but the substance always remains.

In every pebble that lies unheeded on the ground are pent sundry gaseous substances, which only await the delivering hand of the analyser to be liberated and expanded; possessing in their free and etherealised existence many powers and properties which they were debarred from exercising while imprisoned in their condensed and materialised form. To the ordinary observer, the stone thus transmuted in its form appears to be destroyed, but its apparent death is in reality the beginning of a new life, with extended powers and more ethereal substance. Thus it is that physical death acts upon mankind, and in that light it is regarded by the true and brave spirit, with whom to live is toil, and death is a new birth into life, of which he is conscious even here. Death is to such minds the greatest boon that could be conferred upon them, for just as the destruction of the pebble etherealises and expands the element of its being, so by the death or destruction of the body the spirit is liberated from its material prison, and humanity is divinised through death.

And also the following, which I select because it embodies my father's great principle, that scientific phraseology is in place only in strictly scientific works written expressly for strictly scientific readers, and that in books written for the general public it may and must be dispensed with.

The observer can, in a minute fragment of bone, though hardly larger than a midge's wing, read the class of animal of whose framework it once formed a part as decisively as if its former owner were present to claim his property; for each particle of every animal is imbued with the nature of the whole being. The life-character is enshrined in and written upon every sanguine disc that rolls through the veins; is manifested in every fibre and nervelet that gives energy and force to the breathing and active body; and is stereotyped upon each bony atom that forms part of its skeleton framework.

Whoever reads these hieroglyphs rightly is truly a poet and a prophet ; for to him the " valley of dry bones " becomes a vision of death passed away, and a prevision of a resurrection and a life to come.  As he gazes upon the vast multitude of dead, sapless memorials of beings long since perished, " there is a shaking, and the bones come together" once again ; their fleshy clothing is restored to them ; the vital fluid courses through their bodies ; the spirit of life is breathed into them ; "and they live and stand upon their feet.ᵛ  Ages upon ages roll back their tides, and once more the vast reptile epoch reigns on earth.  The huge saurians shake the ground with their heavy tread, wallow in the slimy ooze, or glide sinuous through the waters ; while winged reptiles flap their course through the miasmatic vapours that hang dank and heavy over the marshy world.  As with them, so with us—an inevitable progression towards higher stages of existence, the effete and undeveloped beings passing away to make room for new and loftier and more perfect creations.  What is the volume that has thus recorded the chronicles of an age so long past, and prophecies of so far-distant a future?  Simply a little fragment of mouldering bone, tossed aside contemptuously by the careless labourer as miner's "rubbish."

Not only is the past history of each being written in every particle of which its material frame is constructed, but the past records of the universe to which it belongs, and a prediction of its future.  God can make no one thing that is not universal in its teachings, if we would only be so taught ; if not the fault is with the pupils, not with the Teacher.  He writes His ever-living words in all the works of His hand ; He spreads this ample book before us, always ready to teach, if we will only learn.  We walk in the midst of miracles with closed eyes and stopped ears, dazzled and bewildered with the light, fearful and distrustful of the Word.

It is not enough to accumulate facts as misers gather coins, and then to put them away on our bookshelves, guarded by the bars and bolts of technical phraseology.  As coins, the facts must be circulated, and given to the public for their use.  It is no matter of wonder that the generality of readers recoil from works on the natural sciences, and look upon them as mere collections of tedious names, irksome to read, unmanageable of utterance, and impossible to remember.  Our scientific libraries are filled with facts, dead, hard, dry, and material as the fossil bones that fill the sealed and

F 2

caverned libraries of the past. But true science will breathe life
into that dead mass, and fill the study of zoology with poetry and
spirit.

Such digressions from the main principle of the
work occur not uncommonly throughout the three
volumes, generally at the close of a chapter—if such
the divisions of the book may rightly be termed—where
the leading characteristics of a group of animals are
being summed up, and a few general conclusions drawn.
And in most cases they illustrate one of the leading
principles of his writings—of which he often spoke to
intimate friends, although never formulating it in print
—namely, that in writing books of such a character as
his own, religious instruction, while it should never be
brought obtrusively forward, could and should always
be afforded by implication. More than once, when
writing for magazines of an avowedly religious character, editorial additions were inserted after the proofs
had passed through his hands, generally consisting of
Scriptural quotations which seemed specially applicable
to the subject under treatment. These always made
him furious, and usually resulted in a strong letter of
expostulation; for he was accustomed to say that, while
he always endeavoured to teach religion in all that he
wrote, he never attempted to force it upon his readers,
but always left them to gather it half-unconsciously
from the general tenor of his writings. *O si sic omnes!*

# CHAPTER V.

Appearance of the larger Natural History—"Common Objects of the Microscope"—The "Old and New Testament Histories"—"Glimpses into Petland"—Incredulous Critics—"Homes without Hands"—Review in *The Times* — A curious characteristic — Editorship of *The Boy's Own Magazine*—"The Zoological Gardens"—Failure of the publisher—An amusing correspondence—"Common Shells of the Sea-shore"—"The Fresh and Salt Water Aquarium"—"Our Garden Friends and Foes"—Commencement of "The Natural History of Man"—Preliminary investigations—Collection of savage weapons and implements—"Bible Animals"—How the double work was performed—The *raison d'être* of "Bible Animals"—Its completion and appearance in volume form—"Common British Moths," and "Common British Beetles"—"Insects at Home"—The "Modern Playmate"—"Insects Abroad"—Difficulty of obtaining information.

THE first part of the great Natural History was published in the month of March, 1859, and for forty-eight consecutive months the parts regularly appeared, until the whole animal creation, from the anthropoid apes down to the infusoria and the sponges, had been carefully and systematically described. The book was by no means a strictly scientific work, in the ordinary sense of the term. It was intended for the general public rather than for a special and limited class of readers, and aimed, as all its predecessors from the same pen had done, at making the study of zoology bright and interesting to those who knew little about it, while yet the need for accuracy was carefully kept in mind throughout. In fact, to quote the words of the preface

to the first volume, the work is, and was meant to be,
"rather anecdotal and vital than merely anatomical and
scientific." For my father always held that the object
of the true zoologist is "to search into the essential
nature of every being, to investigate, according to his
individual capacity, the reason why it should have been
placed on earth, and to give his personal service to
his Divine Master in developing that nature in the
best manner and to the fullest extent." And therefore he relegated the whole of the classificatory
portion, consisting of an elaborate compendium of
generic distinctions, to the end of each volume, in
order that it might in no way interfere with the more
popular portion of the work.

This Natural History, however, was not the only
work undertaken during the years 1859—62, for besides
various magazine articles, some of them of no inconsiderable length, the third of the "Common Objects"
Series—"Common Objects of the Microscope"—made
its appearance in 1861. In this little book, however
—almost for the only time in the whole of his career
—my father availed himself to some extent of the services of a *collaborateur.* Not in the actual composition of the book, for he wrote every word himself;
neither was it a mere hasty compilement to suit the
needs of the moment. But the great and incessant
pressure upon his time led him to relegate the selection
of the objects to be described to other hands; and so
this part of the work was entrusted to Mr. Tuffen
West, who employed the greater part of a year in col-

lecting specimens for that special purpose. Messrs.
Baker, also, the well-known opticians of High Holborn,
most liberally placed their entire stock of instruments
and slides at my father's disposal; and so, the mechanical
part of the labour being so greatly lightened, he was
able to write the book in such odd moments as were not
monopolised by the Natural History.

In 1862, as already mentioned, came the resignation
of the chaplaincy at St. Bartholomew's Hospital, and
the removal from London to Belvedere, better known at
that time as Lessness Heath. And the following year
witnessed the appearance of no less than three books.
Two of these, however, were quite of small size, and,
under the title of the "Old and New Testament
Histories," consisted of a short and concise account of
the Scriptural narrative, written in plain and simple
language for the use of children. These two little
books were perhaps the pioneers of Bible manuals for
the young, and met with a tolerably large sale, although,
as was usually the case, my father profited but little by
their success.

The last of the three books was of quite a different
character, and, under the descriptive title of "Glimpses
into Petland," comprised short biographies of a number
of pet animals, nearly all of which had been in the
possession of my father himself. "Pret," the cat, and
"Roughie" and "Apollo," the dogs, together with
chameleons, chicken-tortoises, lizards, and butterflies, all
were described in turn, in manner entirely anecdotal, and
from the point of view of one who regarded them as

intelligent and even rational beings.  The book en-
countered rather merciless treatment from some of the
reviewers, who apparently could not bring themselves
to believe that the stories recounted therein were
true.  But it met, nevertheless, with much favour at
the hands of the public, and, just twenty years later,
was reissued in a revised and extended form.

In 1864 my father began perhaps the most popular
work which he ever wrote, and which has always been
specially associated with his name—the well-known
" Homes without Hands."  In this he set himself to
describe the various habitations constructed by different
animals for the use of themselves or their young, a
task which he completed in a stout octavo volume of
some six hundred and thirty pages.  The work, however, which appeared under the auspices of Messrs.
Longmans, Green, & Co., was in the first place published in monthly parts, just as the larger Natural
History had been; and its publication in volume form
did not take place until 1865.

The popularity of the book was soon assured, even
if the previous issues of the monthly parts had not
paved the way for its production as a whole.  Only
a few days after its appearance *The Times* devoted no
less than four columns to a review of the work, and
spoke of it throughout in the very highest terms.  The
other newspapers, daily and weekly, followed suit, and
the consequence was that, perhaps putting " Common
Objects of the Country " out of the question, " Homes
without Hands " proved by far the most popular

and successful of all the numerous books which proceeded from my father's pen during his thirty years of literary life.

In this work I notice particularly that which was perhaps a characteristic of all his writings, namely, the utter absence of anything whatever in the way of a peroration, or even of a thought-out and carefully turned conclusion. He usually began both his books and magazine articles with a thoughtful introduction, comprising a statement of the subject which he intended to treat, and of the point of view from which he was about to consider it. Of this, in fact, he made a systematic practice, often saying that, after settling upon a title for a book or an article, the hardest part of the work was to find a suitable beginning. And I have frequently known him to expend at least as much time and thought over his prefatory paragraph as over the whole of the remainder of the article. But with regard to a conclusion he rarely seemed to trouble himself at all, and merely adopted the simple plan of leaving off when he had said all that he had to say upon the subject. Thus "Homes without Hands" concludes with the sentence—"As is the case with many of the illustrations to this work, the sketch was taken from nature." That is all; nothing more at all. "Common Objects of the Country," in like manner, ends with a sentence equally simple— "Figure 6 shows the curious Earth-star, chiefly remarkable for its resemblance to the marine star-fish." And so on. And

even in "Trespassers," a book in which, perhaps,
almost more than in any other, he had a definite
idea to work out, and a definite ground to cover,
he abstains in the same marked manner from anything
approaching to a recapitulation of what he has said,
or to a summary of the principal points which he has
brought forward.  The final paragraph is merely as follows:—"Now as to the travelling ants, which are shown
in the illustration; these creatures act, when on the
march, just as soldiers when pushing their way towards a battery; they always keep themselves under
cover, and in a most extraordinary manner, and with
wonderful speed, build covered galleries, under the
shelter of which they can proceed unmolested by the
unwelcome light."

Yet, in his sketch-lectures, he was most careful
with his conclusion, and always had a few sentences
which took one back over the ground that he had
covered, and summed up the teaching of the whole.
So, too, in any speech that he made in public;
the end was as carefully considered as the beginning.
Probably, therefore, the absence of peroration in his
writings was intentional; certainly it was characteristic.

In the same year in which "Homes without
Hands" appeared—1865—my father varied his literary
labours by undertaking the editorship of *The Boy's
Own Magazine*.  To this, for some years past, he had
been contributing a series of papers entitled "The
Zoological Gardens," in which most of the inmates

of that wonderful menagerie were in turn described; as well as various occasional articles upon a variety of subjects dear to the boyish heart. I do not know that he much enjoyed the work of editing, which had too much of the mechanical element in it to satisfy him; and certainly he never referred to it afterwards with any expressions of pleasure. But after the end of twelve months or so the arrangement was suddenly brought to an end by the bankruptcy of the publisher; and the editorship of an "Annual," which he had also taken over, came to an untimely end owing to the same cause.

Some of the correspondence connected with the magazine was very amusing. There was the orthodox column upon the last page, devoted to "Answers to Correspondents," and questions of the most ridiculous character naturally came in by almost every post. Of one in particular my father was very fond of telling. A boy wrote in great tribulation to say that he was exceedingly short for his age, that he had ceased growing, and that his deficiency in stature preyed very much upon his mind and spirits. Could the editor offer any suggestions as to ways and means by which this deficiency might be made good? The editor politely replied, through the ordinary medium of the correspondence column, to the effect that the case was a very sad one, that he deeply regretted his inability to advise upon a subject of so much importance, and that all he could suggest was the daily use of the rack! The answer duly appeared in print, and elicited a further letter from the same correspondent, evidently written in all sober earnest and good faith. He was much obliged to the editor for his kindly advice ; he was anxious to follow that advice without loss of time. *But as he did not know at what shops racks were to be bought, would the editor be so kind as to tell him where he might procure one !*

It is scarcely necessary to state that this second epistle remained unanswered.

Next in order came the fourth of the "Common Objects" Series, "Common Shells of the Sea-shore" being the title of the little work in question. This appeared in 1867, in which year was also issued, as a companion volume, "The Fresh and Salt Water Aquarium," consisting of a reprint of a number of articles which had originally been contributed to the pages of *The Boy's Own Magazine.* Save and except that a chapter was specially devoted to aquarium construction and management, this last might for all practical purposes have been termed "Common Objects of the Fresh and Salt Water" ; for there is scarcely one of the common inmates of pond, stream, or ocean which is not therein described, or at the least mentioned.

In the same year "Our Garden Friends and Foes" saw the light : a book whose title is sufficiently self-explanatory. But the principal business of the year was the commencement of the great "Natural History of Man," comprising an exhaustive account of all the savage races of mankind, with details of their habits, dress, hunting, warfare, and all else that appertains to man in his uncivilised state. My father had long seen that his Natural History was incomplete, and must remain so until the human species had received due attention with the rest of animated creation. He had for many years taken a special interest in ethnology, and had succeeded in making himself more or less intimately acquainted with most of those who could speak with authority upon the subject. He had even brought together the nucleus of that which some few years afterwards was one of the finest collections of savage weapons, dress, ornaments, and implements in the world. And so he was perfectly qualified to take the matter up. He made arrangements with his publisher, therefore, of very much the same nature as those which had governed the production of the larger Natural History. The best artists only were to be engaged; thirty-four parts, or numbers, were to succeed one another regularly at intervals of a month; and the whole, when completed, was to be issued in similar form to that of the great Natural History, to which, in fact, it was intended to serve as a sequel.

This work involved perhaps even more labour than its predecessor, although it consisted of two volumes only. The number of books of travel, &c., which had to be consulted was simply enormous. Whole days had frequently to be passed in the Reading-room of the British Museum. Books to the value of more than forty pounds had to be purchased outright. Travellers, and those with a special knowledge of any savage race,

had to be sought out and applied to for information
on subjects merely touched upon or even altogether
neglected in the books. And then, of course, there
came the constant supervision of the artists—nearly
every one of whose blocks had in some way to be
altered before passing into the hands of the engraver—
the revision of the proof-sheets, the ceaseless search for
sources of further information, and the necessary correspondence which the work involved.

In the three years during which this book was in
progress, my father added very considerably to the
collection of savage weapons, utensils, tools, ornaments,
and articles of costume which he had already been
accumulating for several years. I do not think that
this hobby ever cost him very much money, for, with
the exception of a few odds and ends picked up chiefly
at old curiosity shops in London, he purchased very
few of his trophies, but obtained them partly as gifts,
and partly by exchange, from those with whom the
preparation of the book brought him into contact.
And soon the collection began to attain to really imposing proportions. The walls of the entrance hall,
staircase, and my father's own study were hung so
closely with spears, swords, blow-guns, bows and
arrows, and other weapons of war that scarcely anywhere was there a spare inch of space available. Here
hung a Kaffir cradle, roughly constructed from a. strip
of hide, and carefully and elaborately set with beads.
There was a Kaffir girl's dancing-belt, made of large
seeds, and so constructed as to rattle loudly at ever

movement of the body. In one corner were a number of Dyak clubs; one a most formidable weapon made from the lower part of the stem of a small tree, with the roots trimmed off in such a manner that the stumps formed a radiating mass of sharp spikes, calculated to lacerate the flesh of a foe in the most terrible manner. Close by was a most elaborately carved " toko-toko," or walking-stick, from New Zealand. Then there were two formidable Macquarri whips, used in that singular dance in which each performer presents his leg in turn to be gashed by a blow from the other. There was a Patagonian " bolas," or three-ball lasso, just as it had last been used by the native hunter; quite a quantity of tomahawks; and several of the strange wooden swords, set on either edge with almost innumerable sharks' teeth, for the manufacture of which the inhabitants of the Kingsmill Islands were formerly famous. A small cabinet contained a variety of poisons and poisoned arrows, including a vessel of the famous Wourali, and a roll of arrows presented by Charles Waterton himself, the first traveller who succeeded in bringing the poison home to England. In the same cabinet was the stuffed skin of a hedgehog which had fallen victim to a prick from one of the arrows ; strong testimony to the potency of the venom, as hedgehogs have little objection to arsenic, strychnine, and prussic acid, receive the bite of the viper with perfect equanimity, and are practically poison-proof. In every odd corner were hung ornamented gourds, daggers, bead necklaces, bracelets, anklets, aprons, and small ornaments far too numerous to mention. And, conspicuous among all, hung two genuine Eton birches, which had been surreptitiously carried away beneath the clothing of an unsuspected visitor. Before the book was finished the collection had perhaps but two equals in Great Britain— that at the British Museum, and the famous "Christy" collection. A very large proportion of the illustrations in the Natural History were drawn from the objects which it contained, and at last it became evident enough that, if many more additions were made to it, the walls of the entire house would hardly afford sufficient surface for its display. After the book was finished, however, and its work was done, my father practically ceased to enlarge it, and, a very few years afterwards, the collection was broken up and sold.

Heavy as was the labour thrown upon him, however, by the preparation of the "Natural History of Man," and the necessity for sending in monthly a prescribed quantity of manuscript upon a prescribed day, my father did not at all appear to consider that his energies were wholly occupied, and actually entered into arrangements for the simultaneous publication of "Bible Animals" upon a similar principle. The first of the monthly numbers of the new work appeared in 1869, and thus for many consecutive months a double *quota* of manuscript had regularly to be sent in, a double quantity of proofs revised, and a double number of artists' illustrations to be superintended and corrected.

Only a man of the strongest constitution could have performed the work which the preparation of these two

works involved. The three-mile run immediately before
breakfast was now almost the only regular exercise that
my father allowed himself; and during almost the
whole of the rest of the day, from half-past four or
five o'clock in the morning until nearly eleven o'clock
at night, he was hard at work at his desk. The two
hours' sleep after dinner—regularly from two o'clock
till four—probably prevented him from breaking down
altogether. But perhaps the greatest marvel was that
the character of his work did not seem to suffer at all
from its quantity, and that he could write as brightly
and freshly after a long day at his desk as he could
when that day was just beginning. That he injured
his health by this close application to work can hardly
be doubted. He suffered greatly from sleeplessness at
night, and, had he deferred his rising to a more orthodox
hour, would have gained no real additional repose by
doing so. And probably the ill-health of 1877 and
1878 was caused almost as much by the reaction after
all this incessant labour as by the worry and anxiety of
the time.

The character of "Bible Animals" is evident from
its title ; but the book is by no means a mere description
of the living creatures mentioned by name in the
Scriptures. It aims rather at elucidating some of the
obscurities of the Bible by showing the true meaning
of the many references to various animals ; and it is
intended, in fact, as an aid to Biblical study rather
than as in any way a popular Manual of Natural
History.

G

Contemporary history (says the preface), philology, geography, and ethnology must all be pressed into the service of the true Biblical scholar; and there is yet another science which is to the full as important as either of the others. This is natural history, in its widest sense. The Oriental character of the Scriptural Books causes them to abound with metaphors and symbols, taken from the common life of the time. They embrace the barren, precipitous rocks alternating with the green and fertile valleys, the trees, flowers, and herbage, the creeping things of the earth, the fishes of the sea, the birds of the air, and the beasts which abode with man, or dwelt in the deserts and forests. Unless, therefore, we understand these writings as those understood them for whom they were written, it is evident that we shall misinterpret, instead of rightly comprehending them. Even with secular books of equally ancient date, the right understanding of them would be important; but in the case of the Holy Scriptures it is more than important, and becomes a duty.

The importance of zoology in elucidating the Scriptures cannot be overrated, and without its aid we shall not only miss the point of innumerable passages of the Old and New Testament, but the words of our Lord Himself will either be totally misinterpreted, or, at least, lose the greater part of their significance. The object of the present work is, therefore, to take in its proper succession every creature whose name is given in the Scriptures, and to supply so much of its history as will enable the reader to understand all the passages in which it is mentioned. A general account of each animal will be first given, followed by special explanations (whenever required) of those texts in which pointed reference is made to it, but of which the full force cannot be gathered without a knowledge of natural history.

Messrs. W. F. Keyl, T. W. Wood, and E. A. Smith were the artists engaged to illustrate the book, and every one of the hundred designs bears special reference to some passage of Scripture. The work was completed and issued as a whole early in the year 1871, the

" Natural History of Man " having made its appearance in volume form a short time previously.

Two small handbooks of the " Common Objects " series, " Common British Moths," and " Common British Beetles " were also written and published during the year 1870 ; both appearing under the auspices of Messrs. George Routledge & Co. I think that my father was afterwards sorry that he had ever written these two books. He did so under pressure from the publishers, at a period when he had not the time to devote to his subject; and under any circumstances, moreover, he was scarcely the right man for the task, as he had for nearly twenty years discontinued the active collection of insects, and had never given that minute attention to species and specific differences which is essential to those who write even a popular handbook upon the subject. In consequence, the insects selected for description were not well chosen, and the books—as my father himself afterwards freely admitted—are of but little practical value.

The year 1871 was occupied by the preparation of a much larger and more important work, which, under the title of " Insects at Home," comprises descriptions, life-histories, &c., of a great number of representative British insects. The book was far more carefully prepared than its two smaller predecessors. More time could be devoted to its preparation in the first place, and the author had not constantly to write under the haunting dread of an imperative demand for "copy." The illustrations, too, were provided for in a much more satisfactory manner, and comprised more than seven hundred separate figures. The result was a bulky volume of six hundred and seventy pages, in which all the different orders were treated in turn, of course in a strictly "popular" manner. And an illustrated description of insect anatomy, by way of an introduction, and a few remarks upon setting and preserving insects, as a conclusion, rendered the book as comprehensive as could be desired.

During the same year, my father edited a new book for boys, "The Modern Playmate," for Messrs. Frederick Warne & Co., himself writing the articles upon skating and swimming, and superintending and revising the remainder. In 1872 and 1873 no book appeared bearing his name; but he was nevertheless very busy upon "Insects Abroad," a companion volume to "Insects at Home," treating of exotic instead of British insects, the preparation of which occupied very nearly two years. For, of course, the task was a far harder one than had been involved by the compilation of its predecessor. The insects selected were, in many cases, almost utterly unknown; some had even for the first time to be described; and information concerning them could only be obtained with great difficulty. But the officials in the insect room at the British Museum offered every assistance in their power; the artists threw themselves heart and soul into their work, examining, as stated in the preface, some three thousand drawers of insects, each containing an average of fifty specimens, and sparing no pains to obtain the utmost possible accuracy in the illustrations. And in 1874 the book appeared, forming a larger volume than its predecessor by more than one hundred pages. No less than eight hundred and sixty species were described in all, of which six hundred were figured; and both descriptions and drawings were in every case made from the actual specimens themselves.

# CHAPTER VI.

## LITERARY WORK (*continued*).

As already stated, there had been two years—1872 and 1873—during which no work of any importance had proceeded from my father's pen; but, after the publication of "Insects Abroad" in 1874, several volumes made their appearance in rapid succession.

First came "Trespassers," the leading idea of which was to show the tendency manifested in every large group of animals towards usurping the domain which is usually occupied by other families. Mammals, broadly speaking, are terrestrial. Yet the bats fly in the air like the birds, the whales and the seals live in the water like the fishes, while the mole lives under the earth instead of upon it. Among the birds, again, the penguin, the auks, and the divers are water-trespassers, while the ostrich and the emu do not fly at all, but live upon the ground like most of the mammals. Then, among the reptiles, the crocodiles, alligators, and turtles are aquatic, while the flying dragon skims through the air after the manner of the petaurists and the flying squirrels ; while there are even fishes, and fishes not a few, which at times leave their natural element and trespass for awhile upon the land. All these, and many more, are described in " Trespassers," which not only gives an account of the animals themselves, but explains the wonderful modifications of structure which enable them to " trespass." In this book, in fact, my father first dilated upon that great and most important fact the one and the only master-key to all zoology, that Structure is subservient to and dependent upon Habit ; in other words, that every detail of the bodily form of every animal is more or less modified in accordance with the life which that animal is intended to lead. And upon this rule—one to which there is no exception—he was never afterwards tired of insisting, alike in book, in magazine article, in sketch-lecture, and even in sermon.

Almost contemporary with " Trespassers " was " Out of Doors " ; but this was merely a collection of magazine articles which had from time to time appeared in various periodicals, and which were now deemed worthy of re-publication in book form. The first twelve of these were now arranged according to the seasons of the year, " beginning with a winter of activity, and ending with a winter of repose." The Zoological Gardens upon a bleak January day ; a sand-quarry in winter ; a hunt under the bark of decaying tree-stumps in the early

spring. These were the subjects of the first three
papers, and then came others on a certain small pond,
historically known as "Mrs. Coates' Bath," which was
full of aquatic creatures of all kinds; an English lane
in the height of summer; the wood ant; the green
crab; the stinging jelly-fish; the toad; the "Children
of the New Forest"; a blackberry bush in autumn; and
the repose of Nature. And then followed a few miscellaneous papers, two of them bearing the rather
singular titles of "Turkey and Oysters," and "De
Monstris."

In July of this year my father met with a most
serious accident. He was always somewhat notorious
for the brittleness of his bones, and already, in my own
recollection of him, had broken no less than five ribs:
three by collision with a low post when unwisely
running at some little speed in the darkness, in which
his shortness of sight rendered him practically blind;
and two by walking against a tombstone when taking a
funeral, a gust of wind suddenly lifting up his surplice
and blowing it over his head, so that he was unable to
see the obstacle in his way. And previously to this, as
already mentioned, he had broken his leg, and his arm,
and his collar-bone (twice), and his nose, and had also
suffered various dislocations and lesser injuries. But
the accident which now befell him was more serious
by far than any of these, and he never really recovered
from its effects to the end of his life.

He had promised—we were living at Belvedere at
the time—to take the morning duty on a certain Sunday

in July to oblige a clerical friend in charge of one of the
Woolwich churches; and, anxious to employ his time
to the last possible moment, he arranged to leave only
by the last train on the Saturday night. When the
night came, there was no moon; the sky was covered
with dense clouds; and, as it was summer-time, the
wisdom of the Local Board had decreed that the street
lamps should not be lighted. Also, as was usually the
case when he had a train to catch, my father remembered
some important omission at the last moment, and was
consequently very late in starting. Now, while the
greater part of the village of Belvedere stands on high
ground, the railway station, which· is some three-
quarters of a mile distant, naturally lies down below,
in the valley of the Thames; and the descent is by a
long and steep hill, closely overhung with trees upon
either side. Anxious if possible to catch his train, and
so to save himself from a four-mile walk at midnight,
my father ran down the hill, carrying a carpet-bag
and a walking-stick in his right hand. Just in the
darkest and steepest part of the road a cartload of
manure had been emptied for the supply of a neighbouring garden, and carelessly left upon the pathway. And
over this, which he was quite unable to see in the darkness, my father fell heavily, alighting with all his
weight upon his unfortunate right hand, the fingers of
which were still holding the walking-stick and the
carpet-bag. The consequence was that the second,
third, and fourth fingers were fractured, one of them in
two places; two were dislocated as well; and nearly all

the bones of the palm were also badly broken, the
thumb and first finger alone escaping injury.

At the time, and in the hurry of the moment, my
father was scarcely alive to the extent of the damage
which he had sustained. He knew that bones were
broken, but not that the mischief was so severe. And so,
instead of giving up his journey—instead even of seeing
a medical man, and driving into Woolwich afterwards—
he caught his train and went on; passed the night in
great pain, the wounded hand, of course, swelling to
three times its original size, and somehow managed to
conduct the services next morning, although more than
once in imminent danger of fainting. He even celebrated the Holy Communion; *how*, it is not easy to
imagine. At eleven o'clock he conducted the full
morning service, and preached. And then, at last, he
made up his mind to come home and have his injuries
attended to. Of course, by this time, the condition of
the wounded hand was such that it was almost impossible even to examine it. Only with the greatest
difficulty were the fractured bones set, and the dislocations reduced. And for many weeks it remained in
splints and bandages, while not for nearly a year could
it be used again for writing. Indeed, it never quite
recovered its former steadiness and strength. Nervous
tremors and twitches would suddenly seize it; at times
it would tremble so severely as to be practically useless.
And to the end of his life my father could never write
without steadying the right hand with the left, and
seldom even lift a cup of tea to his lips without employing both hands in the operation.

Of course, an accident so severe told very greatly
upon his literary work. He could not write for many
months—save slowly and laboriously with his left hand
—and he suffered under the additional misfortune of
being utterly unable to dictate to an amanuensis. So
that until the damaged hand recovered something of
its former usefulness, the supply of manuscript practically ceased. Most fortunately he had taken out a
policy in an accident insurance company, and was able
to claim the allowance for total disablement; for it was
not until the following year that he was able to earn
anything by his pen.

As soon as he could again set to work, however, he
began to write "Man and Beast, Here and Hereafter,"
an examination into the question of the mortality or
immortality of the lower animals. The matter was one
which for a very long period had occupied his thoughts;
probably since the time when, in reading for his examination for Holy Orders, he had studied Bishop Butler's
well-known "Analogy of Revealed Religion." For
no less than fifteen years he had actually been accumulating materials for the book. And now he set
himself to arrange his masses of notes and correspondence into something like order, and to write the
book itself.

This he did upon a somewhat singular plan. Himself a firm believer in the immortality of animals, as
many others had been before him, he first entered into
a careful examination of the Scriptural evidence for and
against the proposition, and showed that most of the

passages which are generally supposed to bear upon the
subject are mistranslated, while others are misunder-
stood. And also he pointed out that in other passages
not a few, animal immortality is either distinctly
taught by implication, or else that human immortality
is denied. Then, passing from the theological to the
practical side of the question, he proceeded to show
that animals possess—although, of course, in low degree—most of the faculties generally considered, not
only as peculiar to the human species, but also as
purely spiritual in character; that they are often distinguished by reasoning power of a really high order;
and that even the attributes which we usually deem as
strictly appertaining to the immortal soul are by no
means lacking in the animal world. This he did principally by means of anecdotes, collected only from the
most thoroughly trustworthy quarters, and showing all
these different attributes as manifested in practice. And
then he argued that, although the difference between
man and beast in the spirit world might possibly, and
not improbably, continue as vast as now, no proof whatever could be found to show that animals ceased to exist
at the moment of dissolution, but that, on the contrary,
an overwhelming weight of evidence, both scriptural
and other, seemed to testify to their absolute immortality. And he also showed that the doctrine is by no
means a hard one to those who will put prejudice upon
one side, and examine the whole question carefully and
upon its merits.

Long before writing " Man and Beast," however,

my father had publicly expressed his views upon this subject of animal immortality. To quote his own words :—

Some years ago, when writing my " Common Objects of the Country," I ventured to doubt the truth of the popular belief on this subject, and was rather surprised at the result. Almost every periodical which gave a notice of the book quoted the passage, and, with only one or two exceptions, more or less approved of it. The exceptional cases were those of distinctly religious publications ; and they, of course, brought against me " the beasts that perish."

I was also inundated with letters upon the subject. Many of them were written by persons who had possessed favourite animals, and who cordially welcomed an idea which they had long held in their hearts, but had been afraid to express. Many were from persons who were seriously shocked at the idea that any animal lower than themselves could live after the death of the body.

Some were full of grave rebuke, while others were couched in sarcastic terms.

Two are especially worthy of notice. The one contains twelve pages of closely-written, full-sized letter paper, in which the writer tells me that any one who cherished the hope that animals could live after death was unworthy of his position as a clergyman, ought to be deprived of his university degrees, and expelled from the learned societies to which he belonged. This argument was so unanswerable that I did not venture to reply to it.

The writer of the second letter remarked that, whatever I might say, he would never condescend to share immortality with a cheese-mite. I replied that, in the first place, it was not likely that he would be consulted upon the subject ; and that, in the second place, as he did condescend to share mortality with a good many cheese-mites, there could be no great harm in extending his condescension a step further.

But, no matter whether the writers agreed with me or not ; no matter whether they were sympathetic, severe, or sarcastic, they invariably mentioned " the beasts that perish." Some wished to know how it was possible to get over a passage which had always prevented them from indulging in the hope that the animals which they had loved on earth would have a future life; while others brought forward "the beasts that perish" as a crushing and conclusive argument, of which they evidently supposed me to be entirely ignorant.

The reader will therefore see how important it is that the true meaning of the Hebrew text should be known, and that the Psalmist should not be accredited with putting forward a doctrine to which, whether true or false, he makes no reference whatever.

As regards the main idea of the book, and the conclusions to be drawn from the arguments which he brings forward, I cannot do better than quote my father's own closing remarks :—

In announcing my belief that the lower animals share immortality with man in the next world, as they share mortality in this, I do not claim for them the slightest equality. Man will be man, and beast will be beast, and insect will be insect in the next world as in this. They are living exponents of Divine ideas, as is evident from the Holy Scriptures, and will be wanted to continue in the world of spirits the work which they have begun in the world of matter.

But, though I do not claim for them the slightest equality with man, I do claim for them a higher status in creation than is generally attributed to them : I do claim for them a future life in which they can be compensated for the sufferings which so many of them have to undergo in this world; and I do so chiefly because I am quite sure that most of the cruelties which are perpetrated on the animals are due to the habit of considering them as mere machines, without susceptibilities, without reason, and without the capacity of a future.

Of course, " Man and Beast," when it appeared, gave rise to much discussion, and a host of criticisms appeared in the newspapers and other periodicals ; some very friendly, some very much the reverse. But the prevailing tone was that of ardent sympathy with the principle of the work; and until the end of his life my father was constantly receiving letters warmly thanking him for the book, and embodying further facts and anecdotes for his use in the event of a revised edition.

This I have reason to believe my father had in contemplation during the closing years of his life. His working copy of " Man and Beast " is filled with letters from friends and manuscript notes of his own, and he had clearly been collecting material for additions and improvements. And gummed upon the title-page is a printed extract from some religious magazine which struck him very deeply, and to which he often referred in the course of conversation. It refers to the original Hebrew of the term translated in our version as " living soul."

Certain of King James's translators (it says) . . . have rendered the Hebrew word "nephesh," soul, *when referring to man,* quite literally. The fact that the *same word* is applied to *animals* is covered up, or concealed, to all who are not Hebrew scholars, other words being substituted for it, such as " life," or " creature."

In Genesis i. 30, "To every beast of the earth, and to every fowl of the air, and to everything that creepeth upon the earth wherein there is life "; the Hebrew words are " nephesh chaiyah," *a living soul.* Also in Genesis i. 20, " Let the waters bring forth abundantly the moving creature that hath *life* "; literally, *a living soul.*

Ten times is the Hebrew of "living soul" found in the first nine chapters of Genesis, and only once, when it refers to *man,* is it literally translated. In nine other instances, when it refers to the lower orders of creation, is the fact carefully concealed from the readers of the English version. *In seven of the nine instances it is Jehovah who uses this unorthodox language.*

The italics are my father's own, inserted, in the
copy from which I quote, in the red ink which he
used so freely. Apparently this extract was intended
to serve as the foundation of an additional chapter,
for I find enclosed in the book a large sheet of carefully worked-out notes, in which a reference to the
"nephesh" of the Hebrew has a very prominent
place.

"Man and Beast" was succeeded in the following
year—1876—by "Nature's Teachings," the leading
idea of which is the remarkable frequency with which
most striking anticipations of human inventions may
be found in the world of Nature. Man hits upon
what he considers a perfectly original notion, only
to find that Nature has been before him, and that
the prototype of his discovery has been existing for
countless ages in the world, ever ready to tell its
unmistakable story to those willing and able to
learn. And from this was drawn the corollary that,
in the words of the Preface, "as existing human
inventions have been anticipated by Nature, so it
will surely be found that in Nature lie the proto-
types of inventions not yet revealed to man. The
great discoverers of the future will, therefore, be
those who will look to Nature for art, science, or
mechanics, instead of taking pride in some new
invention, and then finding that it has existed in Nature
for countless centuries."

The first part of the book is Nautical, and shows
how the sail, the rudder, the oar, the screw, the

cable, the anchor, and the boat-hook are all to be
found typified in Nature, together with almost every
other appliance which man employs for navigation in
its widest sense. Part II. treats of War and Hunting, and shows the analogy between the offensive
and defensive weapons of man, civilised and uncivilised, with those in the possession of animals
which existed in the world long before him. The
pitfall, the sword, the spear, the net, the "gin," the
hook, and even the blow-gun, all are shown typically
to be in the possession of divers animals, although
for the most part invented independently by man.
Then follow in turn sections appropriated to archi-
tecture, tools, optics, and useful arts, and the book
closes with a chapter on Acoustics, in which is shown
the analogy existing between the vibration of a violin-
string and that of the notched ridge upon the wing-
case of a cricket; between the trombone and the
throat of the swan; and between our "whispering
galleries" and the natural echo.

The subject had always been a very favourite one
with my father, and he was never tired of working
it out, and showing its manifold applications. He
refers to it in several of his books; he devoted to it
many a magazine article; and few of those who
ever attended his sketch-lectures will fail to recollect
how frequently he brought the same subject forward,
and how he delighted in showing, for instance, the
similarity between our life-boats and the egg-boat
of the gnat; always with the remark that in this,

H

as in numberless other instances, Nature had been beforehand with man. And, alike in book, in article, and in lecture, he always brought the matter forward with the one intention of showing that the same Creator who embodied an idea in visible shape in Nature put the germ of the same idea into the mind of man, and thus that its independent, dual existence forms one of the many proofs of the unity of the Divine scheme. For—as I have already stated —he always preferred, if possible, to point his moral indirectly, and to leave it rather to be drawn by inference than gathered from direct statement. Thus he never seasoned his writings with texts, and seldom even quoted Scripture at all, although he might be enlarging upon the beauties of Nature and the marvels of Creation. Nature to him was God, and Creation was God's work. Yet he always abstained on principle from saying so in so many words, and had the most utter detestation of that form of writing generally described as " goody," which causes many a reader to throw aside many a book in disgust. And yet the whole spirit of his writings breathes the one great truth, which is never expressed but always implied. The " goody " element is carefully excluded, but yet the work which that element is intended to do is done. Throughout all his work the clergyman and the naturalist went hand in hand, and the one was always the exponent and assistant of the other.

# CHAPTER VII.

## LITERARY WORK (*continued*).

AFTER the issue of "Nature's Teachings" in 1876, three whole years elapsed before any further book appeared from my father's pen. For in that year began the great depression in the book trade consequent upon the Russo-Turkish war; a depression which, although fluctuating in its severity, lasted for long after that war had come to an end, and brought about the ruin of thousands. Money was scarce, for no man could venture to say that England would not be drawn into the struggle, and be forced to throw both her arms and her resources into the scale. Everywhere retrenchment was the order of the day, for few could afford luxuries while the national danger seemed so imminent;

and the first luxury which a man denies himself at
times of financial difficulty is the purchase of books.
Firm after firm of publishers failed, and were forced to
make composition with their creditors. Others escaped,
but only by reducing their staff, cutting down expenses
to the lowest possible level, and discontinuing the
production of new books. And so, like many another
writer, my father was forced into practical idleness
between the years 1876 and 1878, his literary labours
consisting merely in the production of occasional articles
for the monthly magazines.

Partly from this cause, and partly, no doubt, from
the reaction consequent upon the severe and incessant
labours of preceding years, his health gave way, and for
many months, while suffering from no definite complaint, he was seriously unwell. If work had come to
him he would not have been able to perform it; and at
one time, indeed, it seemed doubtful whether he would
ever write again. Early in 1878, however, his strong
constitution reasserted itself, and he began to recover.
And from that time to the end of his life he was
scarcely really ill for a day.

In the latter part of 1878 matters began to improve
in the literary world; and in the course of the autumn
of that year he received a commission from the Society
for Promoting Christian Knowledge to write " The
Lane and Field," as one of the series of small volumes
which the Committee were making arrangements to
publish, under the general title of " Natural History
Rambles." Meanwhile the Belvedere house had been

given up (in December, 1876), and after some eighteen months of semi-nomad life, we had settled down at Upper Norwood, quite close to the High Level railway station at the Crystal Palace.

To the palace itself my father soon became a most regular visitor. The library and reading-room attracted him greatly, and thither, during the whole time that we lived at Norwood, he repaired nearly every day, generally for the purpose of making notes, or of working up some subject upon which he was about to write. The music, too, was of course a great attraction. Orchestral concerts were, as they still are, given daily, either in the concert-room or upon the large Handel orchestra in the central transept. A part, at least, of these he nearly always managed to attend, becoming one of a small party of musical enthusiasts, who were looked upon as having a sort of prescriptive right upon ordinary week-days to the seats which the professional critics occupied upon the Saturdays, and who, full orchestral score in hand, were regularly to be seen attentively following the music from the first note of the concert to the last. He was one of the select few, also, who by special favour obtained a card of admission to the rehearsals for the well-known Saturday concerts, of which card he was by no means remiss in making use. And, after his manner, he soon contrived to be on the most friendly terms with every official connected with the building, and to obtain admission at will into those *sacra privata* from which the general public were rigidly excluded.

Before the end of the year "The Lane and Field" had gone to press. It was not a hard book to write, for all the necessary information had been acquired years before, and little now remained to be done save the actual preparation of the manuscript, and the choice of fit and proper illustrations. And so my father was enabled, at the same time, to work at a second book, which for many years he had had in contemplation, but which, for various reasons, he had never until now been able to commence.

This was a new edition of Charles Waterton's famous "Wanderings in South America," which, deeply interesting as it is, is deprived of half its value to the general reader by its author's singular fondness for the use of incomprehensible native titles in place of those familiar to English ears; so that the identification of the various animals, plants, and trees is rendered utterly impossible to those who do not possess a key to this strange phraseology. This key Waterton always steadily declined to provide; and when, after his death, the book was gradually passing out of circulation, it occurred to my father that a new edition, in which an explanatory index should be supplied, while the "Wanderings" themselves remained untouched, might save a really valuable book from falling into oblivion.

And for the preparation of this new edition he was peculiarly qualified. He had regularly corresponded with Waterton; he had visited him at Walton Hall, his marvellous bird-paradise in Yorkshire; and he had received personal and special instruction from him in his wonderful system of taxidermy. And, moreover, ever since first reading the " Wanderings " as a boy, he had neglected no opportunity of identifying the various animals, &c., which are therein mentioned. Such an identification was by no means unnecessary, as my father points out in the preface to the new edition.

The book fascinated me (he says, speaking of the " Wanderings " when first they fell into his hands). Week after week I took it out of the (school) library, and really think that I could have repeated it *verbatim* from beginning to end. . . . But there was one drawback to the full enjoyment and comprehension of the book. It mentioned all kinds of animals, birds, and trees, and I did not know what they were, nor was there anyone who could tell me. I did not know what a Salempenta was, except that it was good to eat. It might be a monkey, a fish, or a fruit. Neither could I identify the Couanacouchi, Labarri, Camoudi, Duraquara, Houtou, or Karabimiti, except that the three first were snakes, and the three last were birds.
It was certainly pleasant to learn that the traveller in Guiana would be awakened by the crowing of the Hannaquoi, but there was no one who could tell me what kind of a bird the Hannaquoi might be. Then, as to trees, I did not know the Siloabàli, or the Wallaba, or even the Purple-heart, nor how the last mentioned tree could be made into a woodskin. I wanted a guide to the " Wanderings," and such a guide I have attempted to supply in the " Explanatory Index."

This " Index," although it consists of less than one hundred and fifty pages, was the fruit of an enormous quantity of labour, the real amount of which did not appear, owing to the fact that it had been distributed over a considerable number of years. The identification

itself was the principal difficulty. Few travellers have
wandered through the wilds and forests of British Guiana,
and those who have done so seem for the most part to
have been somewhat neglectful of its natural history.
All the existing books upon the subject, however, were
consulted in the course of a few days' hard work in the
library of the British Museum; and then came interviews and communications with the officials of the
various natural history departments in the same institution, the leading members of the Zoological Society,
and a number of others who happened to possess special
information upon the subject. And so at length, by dint
of careful and persevering work, all the strange creatures,
plants, and trees mentioned in the " Wanderings " were
successfully identified, and a short account of their life-
history procured for the " Explanatory Index." The
"Wanderings" themselves were left perfectly untouched,
in accordance with the often-expressed wish of their
author. But a short biography of the celebrated
traveller was prefixed to the book, and an account of
his original system of taxidermy written by way of
conclusion; and the book went to press on the last day
of October.

Its passage through the printers' hands, however,
was somewhat delayed; and although towards the end
of November the proofs were coming in daily, it was
not until the 11th of December that the last sheet
appeared, and the index could be drawn up; and the
book had to be on the counters of the booksellers before
Christmas! On that day we had but little time for

rest or refreshment. At half-past seven a messenger from the printer arrived, with instructions not to take his departure until the index was placed in his hands. All the afternoon and all the evening we toiled, and at last, an hour after midnight, the last slip was cut out and gummed in its place, the last reference added, and the boy departed with the manuscript, to get back to town as best he might. And ten days later the first edition of the work was in the hands of the booksellers.

Early in 1879 " The Lane and Field " appeared ; and shortly afterwards my father entered into arrangements with Messrs. Isbister and Co. for the production of a series of six " Natural History Readers " for the use of schools, graduated in accordance with the capabilities of the seven " standards," and prepared in accordance with the requirements of the Educational Commissioners. These books, of course, were to be divided off into "lessons" in the orthodox manner, were to have all words not in common use carefully explained, and were to embrace a review of the entire animal kingdom, from the monkeys down to the sponges. They were, of course, published singly, at intervals, and were quickly introduced into many schools in Great Britain. Recently, too, they have been brought out in America, and have there also met with a considerable amount of success.

A few months later came "The Field Naturalist's Handbook," the joint production of my father and myself, the monthly lists, &c., falling to my share, while he wrote the preface and the introductions to the

various months.  This was followed by " Anecdotal
Natural History," also a joint production, and consisting of a series of articles originally contributed
to the *Practical Teacher*.  Then came " Petland Revisited," a revised, enlarged, and illustrated edition of
the " Glimpses into Petland," which had appeared just
twenty years previously, and which was now brought
up to almost double its original size.  And then my
father set to work upon a book which he had had in
contemplation for many years, and for which, meanwhile, he had gradually been accumulating material.

This was intended to point out the utter absurdity
of the treatment almost invariably received by the
horse.  Man, my father held, with regard to this
animal, sets Nature altogether at defiance.  He creates
for himself an artificial standard of beauty.  He maims
the creature and impairs its usefulness by the very
means which he takes to improve its power of work.
And in all that he does for it, in respect to either its
accommodation, its food, or its personal treatment, there
seems a deliberate wrongheadedness, a careful rejection
of the better way in favour of the worse, which can
only be explained by supposing a sort of inherent
perversity in those who have to do with horse management, and an obstinate determination to travel on in
the same beaten track in opposition to all the dictates
of common sense, of experience, and of science.
Nature permits unshod horses to travel over the
hardest and roughest of ground without the slightest
injury to the hoofs; man, who thinks that he knows very

much better than Nature, considers that a macadamised
road will wear the hoofs away, and so first cuts and
mutilates them, and then "protects" them with a more
or less heavy plate of iron. All the elaborately arranged
system of "springs" in the hoof is utterly destroyed by
our system of shoeing. The foot is rendered liable to
numberless injuries and diseases, and both the strength
and the endurance of the horse are 'greatly diminished,
while experience shows that, even upon the roughest
of roads, unshod horses will do more work, and do it
much better, than those which have been treated—or
rather maltreated—in the orthodox manner.

That is one leading idea of the book. My father
collected information upon the subject from all sources.
He consulted comparative anatomists, professional
farriers, and veterinary surgeons almost without
number. He read every book bearing upon the
subject which he could find. He procured hoofs
and shoes of all sorts and sizes. He induced several
of his friends to take the shoes from off their horses,
and, after a space of three months or so, to allow of
the recovery of the hoofs from the bad treatment
which they had received, to work them as usual, but
without shoes, and to acquaint him with the results.
And when this was done, he made a point, whenever
possible, of seeing the horse for himself, and inspecting
its feet, in order that he might be able to speak and
write about it from personal knowledge. The consequence was that he acquired a vast amount of information upon the subject, and was able to give his judgment

as that of one who thoroughly understood it in all its branches.

But " Horse and Man " by no means treats of the hoof and the shoe alone. That instrument of cruel torture the bearing-rein comes in for strong censure also, and it is pointed out how, in the desire to attain a false and utterly unnatural standard of beauty, the health of the horse is injured, and its capability for work greatly diminished, while it is subjected to positive agony by the abominable appliance in question. Many of the diseases to which horses are subject are traced entirely to its use, while it is shown that it can serve no useful purpose whatever, and cannot be defended on the ground either of utility or of elegance.

Then the ridiculous appliances known as " blinkers " are attacked in their turn; and it is shown that they injure the eye—partly by heating it, and partly by straining the vision—that they are a frequent cause of "shying" upon the part of their wearer, who can obtain merely a transitory and imperfect view of some unfamiliar object, and so naturally takes alarm, while their only possible function is to protect the eyes from the whip of a careless driver! For, as riding horses are never seen with blinkers, it is impossible to pretend that their office is to prevent the animal from catching sight of objects which might alarm it; saddle horses being obviously as liable to alarm as those in harness.

Then follow a few words upon the now, happily, extinct practices of " docking " and " ear-cropping," while the last two chapters are devoted to defects in

our system of stable-construction, feeding, and the causes
of " vice " in horses. And " throughout the work," as
the preface states, "a parallel is drawn between the
horse and the steam-engine, and an attempt is made
to show that those who have the management of the
former or the latter will be adapted to their task in
proportion to their knowledge . . . Engine and man,
in fact, must go together, and so must 'horse and
man.' "

Of course, the book provoked a perfect storm of dis-
approval from professional men, who found their leading
ideas ridiculed, and the many defects of their system
ruthlessly exposed. That my father had expected.
Many of the reviews, too, were adverse, and some spoke
of the folly shown by any one who should take upon
himself to write upon such a subject, not being a
qualified veterinary surgeon. That he had expected
also. So much interested opposition had in the first
instance to be overcome that he did not at all despair
of ultimate success, and he was content to wait quietly
until the new teaching should have had a fair trial,
and its results be made apparent. And by-and-by
letters began to reach him—at first very occasionally,
but afterwards with greater frequency—from those who
had been induced by the book to follow out his instructions, and who, having done so, felt bound to write
and thank him for the results which they had obtained.
Here is one of the most interesting of the letters in
question :—

"PRESTON, 28*th* *April*, 1887.

" DEAR SIR,

"Through reading about ' Unshod Horses ' in your book ' Horse and Man,' my father and I decided to try the experiment ; and on the 3rd or 4th of January of this year we had the shoes taken off one of our ponies (aged seven), and commenced to prepare it for running without them, as advised in your book.  Since the beginning of March it has done its ordinary work without shoes, and I have driven as many as forty miles in the day without any damage to the hoof.

" To-day our man has been down to Lytham, thirteen miles distant, and on the road he was stopped by Inspector T——, of the Royal Society for the Prevention of Cruelty to Animals, and told that we must have the shoes put on again, as in the Inspector's opinion running without shoes was cruelty to animals.

" We feel much inclined to persevere in our course, and should be much obliged if you could give us any assistance or information in the event of the Inspector taking action.

" Awaiting the favour of your reply,

" I am,

" Yours sincerely,

" J—— H. T——.''

This letter—with the exception of the concluding paragraphs—is a type of several which my father received, and of which he was very proud, never letting an opportunity pass of referring to them in his sketch-lectures or his magazine articles.  For he was by no means satisfied with attacking the many abuses of horse management merely in a book, which would, and could, be read only by a very small proportion of the community.  The new teaching, if it were to prosper and to bear fruit, could not be too widely spread ; and so a special sketch-lecture was prepared, and special magazine articles from time to time written upon the subject. The lecture was delivered in all parts of the kingdom, as well as in America; in some cases—as at Bristol—before an audience composed almost entirely of those whom it was specially necessary to influence. Shoes actually taken from the feet of horses were exhibited, together with specimens of hoofs, both shod and unshod. The anatomy of the animal was carefully explained, and illustrated by many drawings in coloured chalks upon the great black canvas. And, with regard to the literary part of the task which he had set himself, almost the very last magazine articles which my father wrote were a series, upon the popular errors of horse management and treatment, for the *Leisure Hour*, which are appearing in that periodical at the time of writing these lines.

In another way, also, he strove to disseminate the leading principles of the new teaching ; namely, by the production of a large coloured chart entitled " Illustrated Stable Maxims," showing by means of coloured diagrams some of the anatomical peculiarities of the horse's body, and intended to be hung in stables for the benefit of grooms and coachmen. The sheet also contains short " Instructions " upon the Structure of the Hoof, Feeding, Watering, Air, the Bearing-rein, Blinkers, and Stable Medicines, and forms, in fact, in the briefest possible form, a kind of epitome of the book.

Almost as soon as " Horse and Man " had been placed in the printer's hands, my father set to work upon another book for the Society for Promoting Christian Knowledge : a more important one than its

predecessor, the "Lane and Field," and treating, not
of natural history at all, in the ordinary acceptation
of the term, but—under the title of "Man and his
Handiwork"—of the gradual advance of the human
race from savagery to civilisation, as shown by the
works of their hands. The book is not a very large
one—consisting of some six hundred and fifty octavo
pages, set in large type, and profusely illustrated—but
includes a good deal. It opens somewhat strangely,
with the sentence "A horse cannot play the piano,"
and then proceeds to show the vast difference of the
human hand from the corresponding member in any
other animal, and the part which it plays in the arts
and inventions, as well as in the daily life of mankind.
And it is also pointed out that the hand of man *makes*
the tools and weapons which, in the members of the
animal world, are provided by the hand of Nature.
Man makes out of iron and wood a spade wherewith to
dig, but the mole, which, in proportion to its dimensions, is a far better excavator than any living man,
has its spade naturally provided for it, in the form of
its own specially-modified fore limbs. Man makes oars
wherewith to row his boat; but the water-beetles and
water-boatmen have their hinder limbs modified into
oars, wherewith they row their own boat-like bodies.
And so on. And then is traced out the gradual development of human tools, weapons, utensils, clothing,
and ornament, from the primitive relics which have
been brought to light by geology, down to their comparatively perfected representatives of the present day

The work, of course, is largely ethnological. It takes in much of the ground already covered in the "Natural History of Man," although approaching it from a totally different point of view, man being considered with regard to what he does, rather than to what he is. And the contrast between civilisation and uncivilisation is constantly brought forward, while at the same time the great gulf fixed, which separates the savage from the brute, is never forgotten.

And it is these two latter points which inspire the closing lines of the book. " I mention this phase of human existence," runs the last paragraph, " as showing the common humanity of ourselves at the present day, and of the races who lived and left their handiwork behind them as the only memorial of a long-vanished epoch. It is impossible, moreover, for anyone to contemplate even the simplest example of man's handiwork, be it but a flint flake or a notched bone, and not to feel that it indicates the impassable gulf which separates the lowest of the human race from the highest of all other inhabitants of earth."

From this it will be seen that my father was no believer in the theory of evolution, at any rate so far as it concerns the development of man. Other passages in his books speak even more strongly and decidedly upon the subject. " Anyone," he says, in the larger " Natural History," with reference to the gorilla, " who could fancy himself to be descended, however remotely, from such a being, is welcome to his ancestry." And for many years after the doctrine was set before the

I

world, I believe, he rejected it *in toto*.  But latterly, I
fancy—although he could never be induced to talk upon
the subject, beyond admitting that, in his opinion, the
theory was in no way opposed to religion—he somewhat modified his convictions, holding that evolution
was no doubt responsible for much in the way of animal
development, without granting to it the almost limitless
power which many naturalists claim.  But it was far
from easy to gather his views regarding a subject upon
which he was so reticent, and upon which he evidently
considered that it was as yet premature to pronounce a
decided opinion.

The last of my father's works published during his
lifetime was the "Handy Natural History," a book
written specially for boys, and produced under the
auspices of the Religious Tract Society.  It treats of
the vertebrate animals only, and has no higher aim
than to give such a pleasant readable account of the
animals which it describes as to lead the youthful
reader to seek other sources of information also, and to
persevere in the study which he has taken up.  The
book appeared in 1886.

Of posthumous works there are three, only two of
which are as yet actually published.  One is "The
Brook and its Banks," originally written in the form of
a series of connected articles for the *Girl's Own
Paper*, which has now, at the time of writing (October,
1889), just appeared in volume form.  It gives a short,
brightly written account of the different creatures,
aquatic and terrestrial, which are to be found in or

about our smaller streams, and contains many references to the zoological experiences of my father's own boy-hood.

Then there is "The Dominion of Man," a larger and more important work, recently published by Messrs. Richard Bentley and Co., under the motto of " The fear of you and the dread of you shall be upon every beast of the earth, and upon every fowl of the air, and upon all the fishes of the sea ; into your hand are they delivered." The book treats of the sovereignty of man over the lower animals, and traces back that sovereignty to pre-historic times, through the medium of the cave and rock inscriptions which have of late so much attracted the attention of scholars. This book was begun and finished in 1887, and is very characteristic both in style and treatment.

Lastly, there is a small book on " Ant Life," which was written in the first place for *The Sun* magazine, and of which the two last chapters were still incomplete at the time of my father's death. The preceding fourteen chapters have already appeared in the magazine in question, and probably will sooner or later be pub-lished in volume form.

# CHAPTER VIII.

## LITERARY WORK (*concluded*).

"Bees and their Management"—"Strange Dwellings"—"The Boys' Own
Natural History"—"Common British Insects"—"Bible Animals" divided
—Reprints—Magazine Articles—Connection with *The Boy's Own Paper*—
*The Sunday Magazine* — "Writings for the Young" — "Mistram"—
Summary of Literary Labours—"Popularising" Natural History—The
"Field" Naturalist—Importance of Classification—Personal Observation—
Opening out a New Field.

BESIDES those which I have already mentioned in the
course of preceding chapters, several books have from
time to time appeared bearing my father's name. Of
these, perhaps, the only one that can claim to take rank
as a separate work is a small handbook on "Bees and
their Management," edited by him in quite the early part
of his career, and published by Messrs. George Routledge
and Co., as one of their series of popular shilling handbooks for the country. He was not very well satisfied
with this himself, I fancy, and afterwards wished that
it had not borne his name. I do not remember ever to
have seen a copy, and have no doubt that it has long
since fallen out of circulation.

The remainder of these books consist partly of smaller
volumes cut down from his larger works, partly of
collections of articles which had been previously published in the periodical press. The first of the former
class to appear, I believe, was "Strange Dwellings,"

extracted, with a few alterations and additions, from " Homes without Hands " ; a book of about half the size, but with sufficient of the original matter to render the scheme precisely similar. The " Boy's Own Natural History," in like manner, was cut down from the large three-volume work which was completed in 1863. " Insects at Home " yielded—in 1882—" Common British Insects," which included a selection only of the beetles, butterflies, and moths described in the larger work. And the list is completed by " Wild Animals of the Bible," " Domestic Animals of the Bible," and " Birds of the Bible," all of which were extracted from " Bible Animals."

Of the second class of these supplementary books, comprising those which consist merely of reprinted magazine articles, " Out of Doors " has already been mentioned as appearing in 1874. This was followed about ten years later, first by " My Back-yard Zoo," and afterwards by " The Romance of Animal Life," and " Half Hours in Field and Forest," all three of which consist of articles originally contributed to *Good Words* and *The Sunday Magazine.* Latterly, also, the Society for Promoting Christian Knowledge has published part of a long series of articles written for *The Child's Pictorial* in two handsome quarto volumes, under the title of " The Zoo." A third volume will probably follow. Their principal feature is that the illustrations—from the pencil of Mr. Harrison Weir—are for the most part in colours. The books, of course, are intended for children's reading.

The magazines to which my father at different times contributed are somewhat numerous; and no doubt to the list which I have been able to draw up many more might be added, of which no record has been preserved.

*London Society* was one of the first for which he wrote. In this appeared four of the papers since reprinted in "Out of Doors," among them "Turkey and Oysters," and an essay entitled "Our River Harvests," giving a description of the science of fish-hatching as it existed in its earliest infancy. In the *Cornhill Magazine* appeared, among others, two long articles upon the inner life of a hospital, written during the time of the chaplaincy at St. Bartholomew's, and one giving a description of Walton Hall, the celebrated abode of Charles Waterton, the traveller, under the title of "The Home of a Naturalist." In the *Boy's Own Magazine*, which flourished—more or less—for about eight years, was published the long series of articles entitled "The Zoological Gardens," which practically amounted to a complete Natural History in themselves before they were brought to a premature conclusion by the bankruptcy of the publisher. And besides these, several other shorter papers appeared in the same periodical upon a variety of subjects not connected with natural history, but dear to the hearts of juvenile readers. In "The Dark Blue" were published "De Monstris"—an account of that very remarkable book the "Nuremberg Chronicle"—"Mrs. Coates' Bath," and several other papers, for the most part treating of some phase of insect life. The *St. James's Magazine* brought out "A Summer Walk through an English Lane," and "The Repose of Nature." *Once a Week* produced "My Toads," "Medusa and her Locks," and several other short essays; and a variety of articles also appeared in *All the Year Round, The Churchman's Magazine, The Home Visitor* (a somewhat lengthy series), and the *Daily Telegraph.*

But during all the earlier part of my father's career he seems to have looked upon writing for magazines or newspapers rather as an occasional relaxation than as an item of serious business. By far the greater part of his time was occupied in the production of books, and his contributions to periodical literature were few and far between. It was not, indeed, until the great depression which existed in the book trade between the years 1876 and 1880 that he set himself systematically to magazine work, with the exception of that period during which he was connected with the *Boy's Own Magazine,* and was writing "The Zoological Gardens" and other papers for its columns. But when no publisher would venture to undertake the risk of producing a book; when those copyrights which he had retained were bringing in little or nothing: then he took to magazine writing upon a somewhat extensive scale. And thenceforward, up to the very end of his life, he was a constant contributor to many magazines, and was generally busy with his type-writer in the production of " copy " for some publication or other.

He was one of the first writers engaged for the

*Boy's Own Paper* when that popular periodical was in contemplation, and contributed an article entitled " Out with a Jack-knife," to the very first number. This was followed by a long series of papers on swimming, skating, and other athletic exercises, as well as on all kinds of subjècts connected with the collection and preservation of insects. After the two first years or so these papers appeared only occasionally, the last which he wrote for the magazine, on Charles Waterton's system of bird-stuffing, being published in—I believe— 1885.

For the companion magazine, the *Girl's Own Paper*, my father wrote only the long series of papers entitled "The Brook and its Banks," already referred to as having been published since his death in volume form. In the *Leisure Hour*, issued by the same society, he wrote frequently, and, as already mentioned, a series of articles upon the treatment and management of horses appeared in its columns after his death.

To the *Sunday Magazine*, however, he perhaps contributed more regularly than to any other periodical, and for many successive years not a volume appeared which did not contain several articles from his pen. Three volumes of these articles—as already stated—have hitherto been issued, and doubtless others will in due course succeed them. To *Good Words*, also, he was a frequent, although not a regular contributor, and was engaged upon a paper to be published in its pages at the very time of his death.

*Longman's Magazine* was another of the periodicals

for which he occasionally wrote, and among the articles which have appeared therein are one upon the whale, another upon the mole, and a third upon " My Garden Wall." The *Church Monthly* has had a series of short papers upon " Our Fellow Creatures "; the *Sun* has published fourteen chapters of the unfinished " Ant " book ; and the *Practical Teacher* brought out twenty-five articles on " Anecdotal Natural History," which have since appeared in volume form, and have also been modified into a series of Natural History Reading-books. For *A 1* he wrote frequently during the first year of its publication, and also contributed at the same time to *Our Own Gazette*, which was then under the same management. *Social Notes*, which appeared for nearly three years under the editorship of the late Mr. Samuel Carter Hall, had many short papers from his pen ; in the *Hereford Times* occasional special articles appeared ; and he also wrote one article each for the *Nineteenth Century* (on " Modern Museums "), the *Argosy*, and *Temple Bar*, the latter of which, I believe, has still to appear.

As a contributor to magazines intended especially for juvenile readers, too, my father was in frequent request, and at different times he wrote for *Little Folks*, *Our Little Darlings* (a long series of papers), and the *Child's Pictorial*, in which appeared the " Zoo " articles now being issued in volume form by the Society for Promoting Christian Knowledge. And, finally, with regard to transatlantic literature, he once contributed an essay (" The Trail of the Sea Serpent ") to the

*Atlantic Monthly,* and was a tolerably frequent contributor during the latter years of his life to the *Youth's Companion,* a well-known American publication.

I may also mention that, about the year 1870, when "Mistram" became for a time a fashionable round game, he published a small pamphlet-handbook upon the game, similar in form to the well-known "Hints upon Whist-playing," by Cavendish. The game was a favourite one with him for a time, and he even managed to work a certain amount of science and calculation into his play.

It now only remains for me briefly to summarise my father's literary work as a whole. It is not for me, of course, to criticise it in any way, or to discuss the question of his position in the dual worlds of science and letters. I may, perhaps, however, be permitted to make one or two remarks upon the labours of his life; and in the first place to point out that he never in any way attempted to pose as a man of science. Quite at the beginning of his career he saw that his *forte* and his opportunity lay in the popularisation of natural history; in making it a study of living and ever-increasing interest to the many, and not a mere monopoly for the few; and in clearing away the hedge of technicalities and repellent phraseology which surrounded it, and which practically debarred all who had not received a special scientific education from taking it up at all. For pure classificators, abstractedly scientific though they might be, he entertained a rooted and measureless contempt. " Certain dreadfully scientific persons," he says in one of his articles, " who call themselves by the name of ' naturalists,' seem to consider zoology and comparative anatomy as convertible terms. When they see a creature new to them they are seized with a burning desire to cut it up, to analyse it, to get it under the microscope, to publish a learned work about it, which no one can read without an expensive Greek lexicon, and to ' put up' its remains in cells and bottles. They delight in an abnormal hœmopophysis; they pin their faith on a pterygoid process; they stake their reputation on the number of tubercles in a second molar tooth ; and they quarrel with each other about a notch on the basisphenoid bone."

His own idea of a naturalist may be seen by what follows :—" Then there are the ' field naturalists,' who delight in penetrating to the homes and haunts of the creatures which they love, and spend whole days and nights in watching their habits. Sometimes a field naturalist remains at home, and immortalises an obscure village by the simple process of using his eyes and telling his friends what he has seen. Another wanders far abroad in quest of new wonders, and if he faithfully narrates the marvels he has witnessed, may calculate on being put down by newspaper critics as a skilful archer with the long-bow."

Not that in any way he failed to recognise the importance of classification, and so of classificators. Such men are a necessity ; for " order is heaven's first law," and without some system of arrangement even the distinctively "field" naturalist would often find himself at fault. But he did from his very heart despise, detest, and abhor the men who pretend to be naturalists and yet have no love at all for Nature, whose only desire upon seeing an animal is the desire to kill it, and who are never thoroughly happy unless they have a scalpel in their hands, and a half-dissected "subject" before them.

He was a compiler himself, no doubt; the accusation has sometimes been brought against him, and the accusation is undoubtedly true. But he was also an original and painstaking observer, and never lost an opportunity of watching an animal for himself, and of recording its ways and doings from personal observation in preference to trusting to the report of another. Such opportunities for observation as he could obtain, he used to the best advantage. He never, so far as I am aware, "read up" a subject simply and solely for the purpose of writing an article upon it; he loved—and showed that he loved—natural history for its own sake; and he would take any trouble in order to clear up a vexed question purely for his own satisfaction, without a thought of its possible usefulness from a pecuniary point of view. And a very considerable proportion of his writings is based entirely upon personal observation.

But what man could possibly do the work which he did without being a compiler? What man, however laborious and painstaking in the field, however diligent

and unremitting in his studies, could hope to produce
such a work as the "Natural History," or even as
"Homes without Hands," from personal observation
and experience? The thing is not to be done. Either
a man must compile, or else such books are not to be
written. But compilation does not necessarily exclude
original work; and when the two·go together in competent hands, a good and useful book is tolerably
sure to be the result.

Undoubtedly my father's great distinction was that
of being the pioneer in the work of popularising natural
history, and presenting it to the general public in the
form of an alluring and deeply interesting study. He
had many subsequent imitators, but he himself imitated
no one. He found zoology a dull and dry study, open
to none but the favoured few who enjoyed special aptitude, special opportunities, and special circumstances for
its pursuit. He left it an open book of world-wide
interest, needing no scholar to ˙read or interpret it, no
unusual zeal or persistence in order to decipher its
secrets. His was the pen that led other pens to write
upon the subject. His was the enthusiasm which fired
the enthusiasm of others; which made observers out of
mechanics, and naturalists out of artisans. And together with ability and enthusiasm he united a dogged
perseverance which enabled him to accomplish a work
which, even so far as its mere extent is concerned, very
few men have excelled.

Is it unbecoming in me, as his son, to say all this?
I trust not. I refrain from expressing any opinion

whatever as to the quality of his work, and I say little of its results; and I endeavour merely to put forward one or two obvious facts, to which I cannot without affectation pretend to be blind, and which, in such a work as the present, could scarcely be passed over without mention. All that I claim for my father is that he did what no man had attempted to do before him; that, alike by pen and by word of mouth, he gave the greater part of his life to the popularisation of natural history; and that in that task he succeeded. Whether he was a man of science, or not, I do not pretend to say; as to whether he was even a remarkable writer, I do not offer an opinion. And if, in what I have said, there be anything which good taste would have left unwritten, I can only deeply regret it, and ask for the indulgence of my readers.

# CHAPTER IX.

## THE SKETCH-LECTURES.

THROUGHOUT almost the whole of his career as an author my father delivered occasional lectures on natural history subjects ; sometimes gratuitously, for the benefit of parochial funds ; sometimes at the special request of scientific societies, and upon the ordinary terms of remuneration. These lectures, as may well be imagined, were very different in character from those which in later years formed an important part of his regular professional work. The power of extempore free-hand drawing had not yet been developed, and he was contented to illustrate his remarks by means of diagrams and specimens. He had not yet struck out the original line which he afterwards made so peculiarly his own. And these, his first lectures, were more orthodox in style than the humorous, unconventional addresses of his later years, for which they paved the way, while wholly unlike them in character.

The first lecture of which I can find any record was delivered in Oxford Town Hall, on March 11, 1856, and was apparently very successful. My father's own entry in his diary for that day is:—

Engaged all afternoon in hanging diagrams in Town Hall. Delivered first part of lecture in evening with much applause.

The preparation for this lecture appears to have occupied some little time, and the lecture itself to have been read from a manuscript. As to its subject, I can find no information.

The first note concerning it occurs on February 1st, when the final arrangements seem to have been made. On the 6th of the same month is the brief entry, "Read for lecture"; on the 8th, "Began to write lecture"; on the 13th, "Made drawings for lecture"; and on the 16th, "Wrote three pages of lecture." After this the entries become more frequent, and there seems to have been trouble with regard to the diagrams. There is an entry to the effect that one of them was "all wrong," and another, on the day preceding that of the lecture, briefly noting that they were "very expensive." Of the second part of the lecture I can find no mention whatever. Possibly it was never delivered at all, as just at this time the chaplaincy at St. Bartholomew's Hospital fell vacant, and on March the 25th my father left Oxford for a short sojourn in town, during which he called upon the members of the Board of Governors. After this, his time seems to have been much taken up with the necessary preparations for removal; and no doubt the second part of the lecture was, in conse-quence, abandoned.

Other lectures were given at Nottingham and at Rotherhithe (on British Guiana) in 1864, at the Stein-way Hall, London, and at Bath, in 1869, and afterwards at Erith, and at Belvedere upon several occasions, of the dates of which I can find no record. But it was not until 1879 that any idea crossed my father's mind of taking up lecturing as a kind of secondary profession. In that year, however, he received a letter from Dr. Chaning-Pearce, who was then residing at Brixton, and who was anxious for a series of afternoon lectures, for the benefit of his friends, in the large geological museum attached to his house. Six lectures were accordingly arranged, admission to be by invitation only ; and the first was delivered on the afternoon of April 12th, the subject being "Hibernation." A large audience assem-bled, and the lecture was enthusiastically received; and on the five following Saturdays the course was con-tinued, with steadily increasing success, the subjects of " Migration," " Aërial Life," " Sub-aquatic Life," and "Subterranean Life " being successively treated, while the last lecture was devoted to such " Unappreciated Insects," as the cockroach, the earwig, the blue-bottle fly, and the gnat. And before it was completed my father had made up his mind to undertake lecturing as

J

a sub-profession, and to devote to it at least the best months of the winter of every year.

A rather amusing incident occurred in connection with one of these Brixton lectures. My father, in the course of his remarks, had happened to speak of the excellence of rats' flesh as an article of diet, pointing out that the almost universal repugnance to it was simply a matter of prejudice, and that it was in reality far more delicate and well-flavoured than that of rabbit or hare, and stating that he himself had often partaken of the dainty in question, and hoped often to do so again. Although the lectures were private, accounts of them were regularly sent to the local newspapers; and by the succeeding Saturday the following notice had been copied into pretty well half the newspapers in the kingdom :—

Having in the course of his lecture to allude to the hedgehog and the squirrel, Mr. Wood observed that it was well known that these animals, when dead and properly cooked, formed excellent articles for human food. Few people were, however, aware that, when similarly treated, the flesh of a rat had a finer flavour, and was altogether a greater delicacy than either of them. There was literally nothing of which he (the lecturer) was so fond as a rat-pie. This was a dish which frequently made its appearance on his table, and was enjoyed by all the members of his family. He had several friends, too, who, like himself, had overcome their prejudice, and thoroughly enjoyed a good helping of rat-pie. He remembered one most interesting case of a whole family who, except the parents, were extremely fond of this dish. They were in very good circumstances, owning large grain stores on the Medway. Their residence was close by, and rats abounded in the neighbourhood. It was always their custom when their parents were out to have their great treat. This being the case on one occasion, dinner-time came, and every one was ready for the repast, which consisted of a pie containing sixteen large rats, when a knock at the door was heard, and in a few minutes grandmamma, accompanied by two young ladies, was announced. Except their savoury pie, the young people had nothing in the house to offer their unexpected guests for dinner, and, what was still more inconvenient, being some distance from a market, nothing could be procured. Undaunted, however, the eldest son, who presided at the table, invited the visitors to sit down to dinner, and, addressing his grandmother, asked whether he should help her to some *gull*-pie. The old lady expressed her astonishment at the idea of their having such a dish for dinner, but at last consented, as did also the young ladies, to take a small piece. This was followed by larger helpings, and, like the rest of the diners, the old lady made a hearty meal. Having come some distance grandmamma stopped the night, and next morning at the breakfast-table, to the great amusement of the children, expressed a wish to have some of the "gull" pie which she had enjoyed so much on the previous day. Inquiries being made below-stairs, it was found that the servants had devoured the fragments. The young lady visitors, too, it was said, were afterwards constantly asking their gentlemen friends, in the season, to shoot them some gulls, so that they might try their hands at making the delicious pie.

Naturally this account aroused a good deal of attention, and my father was shortly the recipient of numberless letters, some simply making inquiries as to the proper way of preparing the novel dainty for table, some asking whether the newspaper reports could possibly be accurate, and some abusing him in the roundest terms for even mentioning the flesh of so repulsive a creature as a possible article of diet.

I quote—*verbatim et literatim*—the most amusing of these last :—

RESPECTED SIR, *April 21st*, 1879.

I had occasion on the 13th of this month to look in our weekly paper and in doing so I noticed your *lecture* on *rat pie* I do

J 2

not know whether you talk for publicity or what but I can never beleive for one moment that you have ever ate rat pie I noticed in your address of the 20th ultmo about hundreds of people writing to you to know how it is made that I believe is a confounded lie for this last two *Sunday Mornings* I have almost been turned sick on observing your lectures on the said *Rat pie* I think the lease you say about the matter the better you are very fond of saying so much about *Rat pie* will you please make this public the next time you lecture that I the undersigned write in the name of hundreds of people to protest against your assertions. All we eat now is not pure so I think you ought not to want us to eat *Vermin.*

I am Sir

A BELEIVER IN HUMAN FOOD.

Please do not forget to make this public with all due respect

Your's truly

.A BELEIVER IN HUMAN FOOD.

The comic papers, too, made much of the opportunity, and for three or four weeks were full of jokes, poems, and illustrations, all bearing on the novel dainty. One of the wittiest of these I give herewith, as for some years my father had it printed, together with the companion illustration, in the syllabus to his sketch lectures. It originally appeared in *Funny Folks* :—

A DAINTY DISH.

The Rev. J. G. Wood, M.A., in the course of a lecture, said there was literally nothing of which he was so fond as *a rat-pie.*"—*Daily Paper.*

On what they like and what they loathe
Mankind are all divided,
And points of taste make woeful waste
Of much for food provided.

This joint we eat, that shun as meat,
    Don't like *this* stew or *that* pie ;
But J. G. Wood says nought 's so good
    In his esteem as *rat*-pie.

With the Chinese he quite agrees
    In relishing the rodent,
And would not wish more toothsome dish
    (Of tooth cook oft might show dent !)
But, though so nice, sheer prejudice
    Makes really " void and null " pie ;
Which taste appals, and so he calls
    His choice confection—*gull*-pie.

This pie to make, a dozen take—
    But stay, the old rule 's binding ;
First catch your rat—and what you're at
    You'll find the need of minding.
Twixt rats of drain, and rats who grain
    Have fed on, frankly, *my* rats
Are barn-floor fed—though Shakespeare said
    The water-ones were pie-rats.*

The rats prepare with special care
    (You wash 'em, draw 'em, flay 'em),
Then, as with pigeon, rabbit, widgeon,
    Within the dish you lay 'em.
Season to taste, put on your paste,
    (For gravy make provision),
Bake to a turn, then serve, and learn
    What luxury 's Elysian !

" But, eat a rat ! "   Just so—it's that
    Which constitutes the shocker ;
As well come with a rat-tat-tat,
    And bid you eat the knocker !

---

* There be land-rats and water-rats : I mean *pirates.*—"*Merchant of Venice.*"

And Wood, M.A., his best may say
  Rodential in laudation,
But the rat-pie will still defy
  All ratiocination.

No doubt these various skits gave amusement to thousands, but they had besides this most beneficial effect, that they made the sketch-lectures known. Those few remarks concerning the flesh of the rat—uttered in parenthesis, and quite without forethought—did what no amount of advertising could have done, for they, or the references made to them, found their way into every newspaper, and everywhere excited public interest and attention. So that when, a few months later, the sketch-lectures proper began, the way was already paved for them, and thousands came to see in the flesh the man who had actually eaten rat-pie!

Before these lectures could be undertaken, however, many preliminaries had of necessity to be arranged. In the first place, an agent had to be found, who should undertake all the correspondence, advertising, and organisation, and leave the lecturer nothing to do but to travel to the appointed spot and deliver his lecture. By a lucky chance my father almost immediately found the very man he wanted, in the person of Mr. George H. Robinson, manager of the New Book Court in the Crystal Palace. Terms were speedily arranged. Lecturer and agent were to share preliminary expenses, and all arrangements were to be made by the latter, who in return was to receive a commission on the fees.

Here, perhaps, I should mention that my father on

no occasion lectured at his own risk. Terms were always made with some local organiser, who paid a definite fee without reference to the receipts, and practically took the entire responsibility into his own hands. By this plan the services of a travelling agent were rendered unnecessary, and travelling expenses correspondingly minimised.

During the ten seasons over which the sketch-lectures extended (1879-1888 inclusive), lecturer and agent worked together upon the most friendly terms, and the arrangement proved a success in every way. Circulars were sent, early in the summer, to all the schools, colleges, natural history societies, and lecture committees in the kingdom, and the dates of lectures so arranged that prolonged tours could be made, with only the minimum of travelling between the different towns *en route.* All correspondence was conducted by Mr. Robinson, arrangements made, fees settled, and even the train service worked out, and the particular trains to and from each town decided upon. Thus weeks before a lecture my father would know when, where, and whence he had to go, the dimensions of the platform, the size of the hall, and the character and situation of the light, and could make his own private arrangements accordingly. All these details both agent and lecturer entered in a huge ledger, with—in my father's case—a kind of abstract—1stly, in his diary, 2ndly, in a pocket note-book, and, 3rdly, in a "memoriser," or folding engagement card.

The next thing was to see about accommodation,

which was generally freely offered. Indeed, throughout his career as a lecturer, my father was hardly ever obliged to sleep at an hotel, owing to the kindness and hospitality of the numberless friends whom he made on his travels; and very commonly he was obliged to refuse three or four invitations, which would perhaps reach him almost simultaneously. And, finally, notices had to be sent round to the different secretaries or managers, with particulars as to time of arrival, luggage, &c., in order that arrangements might be made for conveying both lecturer and apparatus from the railway station to the lecture hall.

Still, however—and before even the preliminary circulars could be sent out—very much had to be done. In the first place a syllabus had to be drawn up; no easy matter, as it practically involved the writing out of tolerably full notes for some fourteen or fifteen lectures, in order that a brief abstract of these might be supplied as a guide to secretaries, &c. Then, when these were printed, a small photograph had to be affixed to each; a practice which was continued until the great increase in the number of circulars sent out rendered it impracticable.

Next a portable drawing-frame had to be designed and constructed; and this proved to be a long and tedious undertaking. My father had soon discovered, not only that he possessed an almost unique talent for what may be best denominated " descriptive freehand drawing," but that audiences were far more interested and pleased by even a rough-and-ready extempore sketch than by the most carefully prepared and elaborate diagrams. He had used the ordinary black drawing-boards for the Brixton lectures, and had found them to answer his purpose fairly well in a comparatively small room. But now, with the prospect before him of lec-turing in large buildings, before audiences of perhaps twelve or fourteen hundred, he clearly saw that some-thing on a larger scale must be provided, if the effect of his drawings was not to be entirely lost. Now what was this "something" to be? Clearly it would be of no use to construct an ordinary blackboard of gigantic dimensions, for the difficulty of carriage would form an insuperable obstacle, while it would be at best but a very clumsy piece of apparatus upon the platform. A blackboard, again, upon the ordinary tripod stand has no sort of stability; a leg slips, or the whole thing is unsteady, or perhaps even collapses with a run. Plainly this would not do at all; but the puzzle was to provide a more efficient substitute.

At length, however, after much thought, and a long discussion with one of the engineers employed by the Crystal Palace Company, the difficulty was overcome. An iron framework of great strength, some seven feet in height by eight in width, was constructed in such a manner as to stand firmly erect when braced up by two strong stays. In the central part of this framework, which was grooved to accommodate them, were placed *four* blackboards, fitting closely side by side, and so accurately adjusted that a perfectly uniform surface was presented when they were tightly screwed together. Ornamental tops to the two upright poles, and three strong boxes to contain the whole, completed the equipment; and so was supplied a blackboard which was large enough to accommodate a drawing eight feet long by four and a half feet wide, which was fairly portable, and which could be easily erected even upon a narrow platform. And now the lecturing began in earnest.

First of all it was deemed advisable, in order to obtain an advertisement for the syllabus, that a course of lectures should be delivered at the Polytechnic Institution, an engagement at which might be looked upon as a sort of " hall-mark " to a lecturer. The prosperity of the building, it is true, was then on the wane, and it did not survive in its original form for much more than a year afterwards. But still it retained much of its old prestige, and certainly an engagement there, for the sake of the advertisement, was not to be despised. Overtures were therefore made, and an engagement for a course of twelve lectures quickly followed; these twelve to consist of the six shortly before delivered at Brixton Rise, and now to be twice repeated, and on no account to occupy more than three-quarters of an hour apiece.

This last restriction chafed my father terribly, for of all things he detested being tied for time, and liked to say what he had to say at what length he pleased, without the necessity for constant reference to his watch. However, there was nothing for it. The programme was made out, and as fast as one entertainment was over another had to be provided; and of course punctuality was an absolute necessity to all concerned.

So the frame was put together behind the scenes, ready
to be carried on and fastened at a moment's notice,
and at the appointed time the curtain rose, and at
the appointed time it fell. The lectures were duly
delivered, the audiences were most appreciative, and
the officials ready and willing to render all the assistance possible. So that when, after the last lecture, we
packed up the frame and drove away, we left the
building with quite a feeling of regret.

Meanwhile, however, other lectures had been going
on at the Crystal Palace; but these need a few words
of introduction.

On the 18th of August my father had received a
sudden visit from two of the leading Crystal Palace
officials (we were then living at Norwood). A baby
gorilla, it appeared, had been safely brought to England, had been secured for exhibition at the Palace, and,
together with a young chimpanzee to which it was
passionately attached, was to arrive at the Palace that
evening. Would my father, as a lover of animals,
like to be present upon the occasion, and supervise the
removal of the animal from the railway station to its
temporary quarters within the building?

Of course he was only too delighted at having the
opportunity of welcoming a stranger of such rarity
(two living gorillas only having previously been brought
to England), and at about a quarter before ten he and
I set out, escorted by a number of officials and workmen,
through the dark, silent Palace, looking very grim and
ghostly by the light of the lanterns, down to the Low

Level Station, to meet the animal. In due course it
arrived, securely fastened up in a crate of portentous
size, and was quickly placed upon a truck, and wheeled
away out of the draught to a warm room for the night.
As soon as it was safely installed, the wrappings were
carefully removed from the crate, and we all peered
in, to see a frightened little black object hiding away
in a corner, and jealously guarded by the chimpanzee.
This was the gorilla; not at all the terrible-looking
creature that one had been led to expect. But then
it was very young—quite a baby, in fact, scarcely two
years old; and it had been travelling in a noisy train
all day long, and it was very tired, and it was very
much dazzled by the light suddenly flashing in upon
it from the lanterns. So a little food and drink were
put into the cage, and the crate was carefully covered
up again, in order to keep out the least suspicion of a
draught. And then we all went home again until the
morning.

Next day we were early at the Palace to ascertain
whether the little gorilla had suffered from its long
journey. There we met the owner of the animal, who
at once suggested that if my father would give a short
descriptive lecture upon the anthropoid apes two or
three times a day it would add greatly to the attractive power of the animal towards the general public.
To this proposition my father, after some deliberation,
agreed. A few specimens, consisting chiefly of casts of
the feet, hands, and heads of adult gorillas, were procured, and between the 21st and 31st of August,

both days inclusive, twenty of these *quasi*-lectures were delivered, making, with those at the Polytechnic, twenty-nine in the nine working days. Of course this involved a good deal of hurried travelling backwards and forwards, from the Crystal Palace to the Polytechnic, and from the Polytechnic to the Crystal Palace ; but the difficulties were overcome, and all the lectures were duly delivered. The health of the poor little gorilla, however, fell off very rapidly. With the seeds of pulmonary disease already working within it when it reached England, it received its final death-blow before it had been in the Palace a week, owing principally to the negligence of its keeper, who was an ignorant Dutchman, with no experience of the larger apes ; for a sudden shower of cold rain, falling upon the heated glass of the Palace, and then rapidly evaporating, quickly lowered the temperature of the building by no less than ten degrees. No precautions whatever were taken to protect the gorilla, or to increase the warmth of its cage, and the animal sustained a severe chill, which, settling upon its lungs, carried it off upon the 3rd of September. On the following day my father assisted at a *post-mortem* examination of its body, and I find in his diary the entry :—

Examined interior of gorilla. *Such* a chest and back. Right lung all tubercles, left very bad.

A second note bears date a few days later.

Chimpanzee died a few hours after leaving Crystal Palace, and an Orang-Outan (not exhibited) died shortly after arrival. But all three anthropoid types were alive at the same time under the same roof, for the first time in the history of the world.

The next lectures were at Forest School, Walthamstow, of which Dr. G. F. Barlow Guy, a very old friend of my father's, was then head-master. The same series were delivered here as at Brixton five months earlier, one lecture being given each day from September 10th to 16th. This was rather an experiment, it being the first time that my father had lectured to an audience consisting almost entirely of boys; but they seemed wonderfully interested, were vociferous in their applause, and were full of questions at the close of each lecture. My father, indeed, always had a wonderful knack of winning and keeping the attention of boys. He seemed to know by a sort of instinct the points which would interest them, and always contrived to put those points in simple and yet attractive language ; so that he never went beyond the capacity of his hearers, and always left them with a clear understanding of that which he had endeavoured to teach. School engagements were consequently numerous, and in course of time a special syllabus was drawn up for schools only, embracing a comprehensive series of lectures on every branch of zoology.

The first lecture of the regular season took place at Lancaster, on September 25th, and was enthusiastically received, the subject being "Unappreciated Insects." The same lecture was given at Swansea, on October 2nd, before an audience of fully eleven hundred, on October 7th at the Angell Town Institute, Brixton, and on the

21st at Newbury, in each case meeting with great success.

Judging by a note in the ledger, dated October 21st, my father does not seem to have carried away a very high opinion of the latter town. " Utterly dull and stupid place," he writes ; " no one seemed to know or care about anything. Partly private room in Commercial Hotel : bad fire ; draughts. Had to hail omnibus for myself."

Three days later followed the same lecture at Maidstone, of the large Natural History and Philosophical Society at which town my father was for many years president. On the 29th he was at Newport-Pagnell, still with the same subject ; and on November 3rd he gave " Life Underground " at the Bow and Bromley Institute. Then followed a short tour, beginning with Angell Town (2nd lecture), on November 19th, and including lectures at Chelmsford, Huntingdon, Cambridge, Huddersfield, Saltaire, Sheffield, and Gloucester (two). At Chelmsford he was greatly struck by the extreme enterprise of a local reporter, who interviewed him in the private room as soon as the doors were opened, wrote an account of the lecture before it was delivered, and actually had it printed before the lecturer had left the platform ! The second of the two lectures at Gloucester was delivered in Dr. Needham's private lunatic asylum, and was listened to by the patients with great interest and atention.

Before the end of the year further lectures were given at Gosport, Bexley, and the Royal Naval School,

New Cross (two); and then came the usual "recess" during the Christmas holidays, when lectures give place a while to pantomimes, and the general public prefers amusement to instruction. A single lecture, however, was delivered at Warminster, on January 13th, but unfortunately a severe snowstorm was raging, and naturally the audience was a very scanty one. On the 20th began a short tour, including lectures at Rotherham, Hull, and Great Grimsby, the latter of which towns is described in the diary as "Queer place, rather of mushroom character; broken up into small cliques, mostly polemic." A lecture at Upper Norwood followed, and then came a week's visit to Bristol, during which five lectures were given.

These were delivered under the auspices of the Royal Society for the Prevention of Cruelty to Animals, and one, to carters only, was of special interest, the subject being "The Horse," and the various mistakes commonly made in the treatment of that animal being carefully explained. Then followed lectures at Plymouth (two), Banbury, Birmingham (two), Gloucester (two), Cheltenham, Greenwich, the Birkbeck Institute in London, and Streatham (five), and the season closed with "Bee Life" at Tonbridge, on June 23rd.

Eighty-three lectures had been delivered in all, including those at the Polytechnic and the Crystal Palace; the result had always been highly satisfactory; and in nearly every case an engagement for the following season was arranged for before my father left the lecture hall. There could be no doubt whatever that the sketch-lectures were a success. People were pleased with the novelty of the thing, with the absence of the usual manuscript and glass of water, and with the rapid impromptu sketches with which the various " points " in the lectures were illustrated. And both lecturer and agent looked forward to a busy season in the following autumn and winter.

Before this could be begun, however, certain alterations had to be made, more especially in the drawing-frame, which had not answered its purpose nearly so well as had been expected. But a notice of these must be deferred to the following chapter.

K

# CHAPTER X.

## THE SKETCH-LECTURES (*continued*).

ALTHOUGH the large portable blackboard which my father had had specially made for these lectures had proved satisfactory in many ways, he soon discovered that, like all experiments, it was open to a great deal of improvement. It was far too heavy, for one thing, weighing no less than 178 lb. when packed in its three boxes ; so that excess railway fares and extra cab hires formed a very important item in the travelling expenses. And it was often difficult, when arriving at a hall, to find helpers sufficiently strong to carry the boxes from the cab to the platform.

Then, again, in spite of its quite unusual dimensions, the board was far too small ! My father had taken to drawing whales and other sea-monsters upon a large scale, and found that a board only eight feet long did not afford him space enough for a sketch of sufficient magnitude; while, even with drawings of lesser size, the available space became so rapidly filled that the constant use of the sponge and towel was necessary. And, thirdly, he wanted a more yielding, more elastic surface. He had now taken to drawing with pastils of many colours—some manufactured specially for himself —instead of with the plain white chalks as at first. The board did not "take" these colours well; and the pastils, moreover, were for the most part so soft that the pressúre against the unyielding wood crumbled them into fragments, so that they could not be used with any degree of certainty. So the big blackboard was doomed.

But what was to be substituted for it? Mr. Waterhouse Hawkins, who about this time had been lecturing at the Crystal Palace, had employed a large sheet of black canvas, loosely stretched by means of guy-ropes; but this did not at all fall in with my father's fastidious requirements, and certainly was rather too much of the "rough-and-ready" description for use upon a public platform. So he devoted himself heart and soul to the designing of a more worthy substitute.

One or two failures, of course, were inevitable. He soon saw that the canvas must be stretched on a wooden frame, but fell into the error of making this frame too light; so that when the strain of the canvas was thrown upon it, it collapsed, and proved utterly unsuitable to its purpose. After quite a long succession of experiments,

K 2

therefore, a second frame was constructed of a far stouter character, and this, after a few alterations and improvements had been made, proved perfectly satisfactory, and afforded a clear surface of eleven feet by five feet six inches ; so that a drawing made thereupon was clearly visible in every part of the largest hall.

Few of those who have seen this great black screen erected upon a lecture platform would be likely to form any idea of the intricacy of the mechanism ; and so perhaps a short description of the apparatus may not be out of place.

In the first place, each of the two upright posts to which the oblong frame actually bearing the canvas was affixed was composed of two pieces, jointing in with one another, and firmly fastened together by long hand-screws. The oblong frame itself consisted of six parts, two uprights being fastened by screws to the outer poles, and four cross-pieces, capable of being firmly braced together so as to form but two, running from corner to corner of these so as to form the oblong. A stout iron brace in the centre of this completed the actual framework, which was supported by four long guy-ropes running from pegs at the top of the outside poles.

The canvas was quite a work of art in itself. In the first place, it had to be of a special quality, finely and yet very strongly woven, and, of course, without seam. Then the edges had to be strengthened with broad bands of the very strongest webbing, sewn on by machine, and running completely round the canvas. Next a number of iron rings had to be fastened to the edges, at intervals of about five inches, by means of twine; and, finally, these rings had to be fitted with running cords carefully waxed, spliced, and bound, to prevent any possible chance of fraying.

Then, of course, this canvas had to be painted; but before that operation could be performed it was necessary to stretch it upon the frame. This was done as follows :—

First, the two upper corners of the canvas were firmly lashed, by means of short cords depending from the corner rings, to the hand-screws upon the upper part of the back of the frame. Next the running cords already mentioned were looped over a number of projecting screws with which the whole inner edge of the framework was set; and, when these cords were drawn tight, the result was that the canvas was stretched so perfectly that not a wrinkle was anywhere to be seen. Then all the cords were securely fastened, and tucked away out of sight, a final turn given to all the screws, and the guy-ropes tightened up if necessary; and then the frame was all ready for the lecture.

The painting of the canvas was a somewhat lengthy process. First of all it had to be " sized," and then left for a day. After this three coats of black paint were successively applied at intervals of a couple of days ; and then was superimposed a coat, sometimes repeated, of "flatting," to employ the technical term, with which a quantity of emery powder had been mixed. The object of this, of course, was to give a certain amount of roughness to the face of the canvas, and so to enable it to hold the chalk. After this "flatting" had dried the canvas was ready for use, and would generally last for a whole season without needing repair; and when it began to show signs of wear and tear, a fresh coat of flatting was generally all that was required.

When not in use this frame reposed in a strong canvas case, protected by stout cushions at the ends, and fastened by means of a rather complicated system of lacing. It was a very good case, and did its work well; but the worst of it was that it looked so dreadfully suggestive of a corpse. Even the railway authorities noticed this, and so striking was the resemblance that, shortly after the mysterious disappearance of the body of the late Lord Crawford, my father was actually stopped upon one occasion by the officials, and compelled to open his great black package before they could be induced to believe that the body of the missing nobleman was not reposing therein. Ever-after this, of course, the frame went by the title of "Lord Crawford," and so it was known upon almost every railway in the kingdom. And there were few of the larger stations, at any rate, at which all the porters were not thoroughly acquainted with the familiar black parcel.

This frame was entirely my father's own invention, and very proud of it he was; so much so, indeed, that he always objected to more help than was absolutely necessary in putting it up, and would trust no one but himself to stretch the canvas, fasten the guy-ropes, and give the finishing touches. But the strain must often have been a severe one, especially when, as often happened, he had been travelling all day long, and only arrived just in time to make the necessary preparations before delivering his lecture. Even under the most favourable circumstances the screen could never be erected in less than an hour, and the physical exertion involved, particularly in stretching the canvas, was very severe, as I can testify from much personal experience. And very often it would happen that the hall-keeper was unable, or unwilling, to give the necessary assist-ance. Hall-keepers, in fact, were a source of constant trouble to my father, and he often inveighed bitterly against them—as a class, that is, for he met with many shining exceptions. One great difficulty which he often had with them was caused by their reluctance to allow him to put the necessary screws into the floor. In the case of the old cumbersome blackboard, there certainly was some excuse for their unwillingness, for the screws used were the well-known " stage-screws," which make a ragged hole in the boarding fully half an inch across But when the new frame took its place these were re-placed by ordinary screw-eyes, making holes so small as to be barely noticeable ; and, indeed, in putting up the frame for a second lecture, it often required a very close search to find the old holes.

But some of these hall-keepers were as jealously careful of their floor as a yacht-owner of his deck, and

a great deal of opposition had to be overcome before
even these small screws could be inserted. In one or
two extreme cases, indeed, my father was compelled to
inform the obdurate janitor that unless the screws were
at once inserted he should decline to deliver his lecture,
on the ground that his drawing-frame could not be
erected, and should lay the matter before the committee—
an expedient which never failed to produce the desired
result.

Taking down and packing the frame after the lecture
was over was also a rather long process, and was seldom
completed much under three-quarters of an hour. First
the canvas had to be thoroughly washed and dried,
and this always took some time, as coloured chalks
have a way of obstinately clinging to the canvas, and
cannot be removed without some little trouble. Then
the frame-work had to be taken to pieces and carefully
packed together; the screws put into one bag; the
cords rolled up and placed in another ; the canvas carefully and neatly folded, and placed on the top of all ;
and then, after several straps had been tightly drawn
round them, the whole had to be transferred to the long
leather case, and securely laced up therein. So that,
with a lecture occupying fully an hour and a half,
visitors to be interviewed, and their questions answered,
perhaps a reporter to be posted up in some part of the
lecture which he had missed, and then the screen to be
taken down and packed, it was often close upon eleven
o'clock before my father could leave the hall. And as,
very often, he would be obliged to take his departure

not later than seven o'clock on the following morning
in order to fulfil another engagement, and would constantly write articles, &c., while in the train, it may
well be imagined that these prolonged tours were a great
strain upon him, and must have told not inconsiderably
upon his general health.

While at home he was constantly going over his
frames (for he had three in all) in order to provide
against accidents.     Perhaps a new canvas had to be
made, or a few rings fastened on more securely; or part
of the woodwork needed repair, owing to rough treatment received upon the railway ; or some of the screws
would be "stripped," and so would have to be replaced
by others.     Perhaps, too, some new improvement had
to be made, involving a visit to the blacksmith, or the
purchase of a new set of screws.     This was commonly
the case; and when, a few weeks after my father's death,
I had to put up the frame for a lecture of my own, I
found that it had changed in several respects during
the year that had passed since I had seen it.

And even if nothing in the way of repairs or improvements happened to be necessary, still almost my
father's first proceeding upon reaching home was to put
up one of his frames in a large lumber-room, either for
the purpose of fitting a new canvas, in readiness for
future necessities, or else that he might practise some of
his drawings.     For all those wonderful sketches, produced so rapidly before the eyes of the audience, and
seemingly without a moment's consideration, were the
outcome of long and careful prior preparation.     First he

used to make a tracing, if possible, of the object he
wished to draw from some thoroughly trustworthy wood-
cut. Then he would copy this two or three times upon
a slate, which hung by a cord from his table, always
attempting to do so with the fewest possible lines, and,
as he frequently used to say, "making every line tell
its own story." Then, having contrived this to his own
satisfaction, he would make a very careful sketch *in
colour* upon the back of one of the small paper strips
which contained his brief lecture-notes. And finally,
chalks in hand, he would go off to his frame and
practise that drawing diligently, until he could exe-
cute it accurately without hesitation and without a
mistake.

No doubt these coloured sketches contributed more
than all else to the invariable success of the lectures.
Every drawing elicited a round of applause, and the
newspapers always commented admiringly upon the
great artistic power which could produce such a result
with such simple means, and apparently with such
perfect ease. In the words of Dr. Oliver Wendell
Holmes, who was present at one of his lectures upon
"Pond and Stream" during his first American tour
and who afterwards wrote to him a letter of warm
admiration :—

I looked as well as listened, and saw the stickleback and his
little aquatic neighbours grow up on the black canvas from a mere
outline to perfect creatures, resplendent in their many-coloured
uniforms. The lecture had much that was agreeable, but the coloured
chalk improvisation was fascinating to the old and young alike, and
was—as it deserved to be—heartily applauded.

I also quote a few lines upon the same subject from the *Marlburian*—the private journal of Marlborough College—which appeared shortly after the first lecture had been delivered there :—

His appliances for illustration were excellent, and the result of no little experience. A capacious canvas, which could be withdrawn into a wondrously small compass ; chalks of all colours, selected to rescue the entangling lines of insect-organs from the confusion inevitable if portrayed in academic white alone; and these, too, bought at great price, and suited for use in the glare of gas-light— such light turns white to yellow ; yellow must be used for white ; blue looks green, and green blue. That red which glowed on the wings of a new-born gnat was hunted up and finally captured only in Paris. Those who have tried to draw under such conditions at close quarters will appreciate, too, the facility displayed in the use of these, the truth to Nature, as far as is possible in section drawing, and the accurate knowledge of detail, which only comes of careful out-door study the motive of which is love.

The first drawing was a longitudinal section of a type-insect. An impossible monster, therefore, but designed to show the structure and position of the internal organs, and necessary to explain the deeply esoteric title of the lecture. If it be not profanity to divulge in print what we were told would not bear publication, that title was "Entomarchetype." The prodigy grew on that magic canvas for all to behold, and though one felt it hard to like a beast with such a name, it was not without a pang that we saw it make way for more familiar creatures under the visitation of that remorseless sponge.

I may perhaps also be permitted to quote the following from the *Altrincham and Bowdon Guardian* of October 8th, 1881 :—

Mr. Wood's method of lecturing is, we believe, unique. It consists in producing upon a black canvas screen drawings of the objects to be described. These are drawn in the presence of the audience

They are not mere diagrams, but finished pictures in colours of great beauty. These, as they literally started into life under the lecturer's artistic touch, elicited very marked approbation from the audience. One picture especially showed the very highest skill. A particular species of the hydrozoa had to be described, which, from the transparency of its substance and the near approach of its refractive power to that of water, can scarcely be distinguished from the element in which it swims. It requires a practised eye and close attention to see it at all. Mr. Wood drew it on the screen as one would gradually come to distinguish its parts ; here a flash of light and there a filament; here a red, and there a blue tint, till the creature ultimately took shape and stood forth in all its beauty of iridescent colours and gracefulness of form.

Of course, all this artistic skill was the outcome, not merely of much careful practice at home, but also of many experiments and failures with chalks of various descriptions. At first the ordinary drawing pastils were procured, but these were soon found to be utterly unsuited for the purpose, having neither sufficient brilliancy of colour to show out upon the black surface, nor the peculiar quality necessary for ready adherence to the smooth canvas. In the course of a conversation, however, with the late Mr. Waterhouse Hawkins, who had already adopted the coloured pastils in his lecture-sketches, although upon quite a small scale, my father was recommended by him to pay a visit to Messrs. Lechertier & Barbe, of Regent Street ; and from that firm he thenceforth procured the whole of his pastils. Some of these, as stated in the *Marlburian*, were imported from Paris, more particularly the brilliant scarlet which glowed out with such striking effect upon the black of the canvas. Some were specially manu-

factured for him, upon condition of his- purchasing a
sufficient quantity to ensure a profit upon the whole.
And a few, but very few—were selected from those
already in stock. By degrees, he added considerably to
the number of different shades which he employed, and
in his boxes I find three shades of red, three of blue,
two of green, three of yellow, an ochre, a brown, and a
neutral tint, all of which have clearly been used upon
several occasions.

These he used with wonderful discrimination and
judgment, seeming to know by a kind of instinct just
how and where to apply the colours so that they might
produce the desired result at a distance. He had,
indeed, something of the peculiar art of the scene-
painter, whose productions, when viewed at close
quarters, appear but the coarsest and clumsiest daubs,
but when seen from some thirty or forty feet away are
really elegant pictures. As one stood upon the platform, quite close to the screen, it was sometimes almost
impossible to realise that these were the famous drawings, which had made the success of the sketch-lectures.
But, upon going down to the middle of the hall, and
then viewing the self-same sketches, one's wonder ceased,
for now one saw them as they were intended to be seen,
and as they were seen in the mind's eye of the lecturer.

This aptitude for drawing was quite a natural gift,
for my father never had a drawing-lesson in his life, and
was entirely his own pupil. During his university
career he showed a good deal of artistic talent in the
production of a number of small pen-and-ink sketches

which now adorn a small scrap-book in my possession. These are chiefly of a humorous character, and for the most part relate to scenes of collegiate life, the key to which, of course, is now missing. But that he added to these in after life is evidenced by the appearance of " Ye Margate Mag-Pyes," a fancy sketch of two elderly gossips conversing upon the Margate sands, and by a whole series of illustrations of pelicans and flamingoes in eccentric attitudes, drawn from life at the Zoological Gardens, and reproduced almost exactly in the small sketches of those birds which appear in the " Explanatory Index" to his edition of " Waterton's Wanderings." He would occasionally illustrate letters to intimate friends, also, by small drawings, sometimes of a wonderfully graphic character. But he never pretended to be an artist, and, indeed, always disclaimed the title, saying that he could show what he wanted to show by a rough drawing, but that anything in the shape of a finished sketch was altogether beyond his powers.

In the later years of his sketch-lectures he was very particular about light, and would have neither gas nor lamp anywhere near the screen itself, save in the form of footlights some six or seven feet in front. He even had a set of footlights made, and for some time carried them about with him. But, finding that they were seldom suited to the requirements of any particular room, he contented himself with sending notice that footlights must be provided in some form or other, and that, if nothing better were forthcoming, half-a-dozen

ordinary paraffin lamps, such as those made to hang
upon a kitchen wall, would answer his purpose. For,
in order that the drawings should show out to the best
advantage, it was necessary that the light should come
from *below.* If it proceeded from above it merely
dazzled the eyes of the audience when reflected back
from the screen, and prevented them from seeing the
sketches at all, while side-lights were almost equally unsatisfactory, and so were tabooed also. But light from
below brought out the full effect of the colours, and
showed them out in bold·relief with the dull black
of the canvas; and if there were no other light in
the hall at all, the result was even more satisfactory
still.

Of course some of his sketches were more striking
and remarkable than others. One of his best was that
of two ants fighting, in which jaws, limbs, and antennæ
were hopelessly interlocked, and yet the individuality
of each insect was clearly preserved. There was a
drawing of the spermaceti whale, too, in which the spine
came first, and then was followed by some of the other
bones and internal organs, while, finally, a line was
quickly run round these and the whale seen to be complete, with every part in due proportion. And the
drawing of the male stickleback in all the glories of his
courting array always elicited a special round of applause. The odd thing was, that no line was ever
rubbed out, no alteration ever made. The sketches
were hastily executed, but were always perfectly exact
in every particular. No measurements seemed to be

taken, and yet the proportions were invariably correct. Of course there was a great deal of art in this, although it did not appear—*ars est celare artem*—and any one who thinks otherwise has only to try to reproduce the drawings, even with the small coloured sketches to guide him, in order to find out his mistake.

# CHAPTER XI.

## THE SKETCH-LECTURES (*continued*).

WITH all the advantages of the new canvas drawing-screen, and a complete set of coloured chalks, the second season of the sketch-lectures began on September 15th, 1880, again with a short series at Forest School, Walthamstow. A completely new course of subjects was chosen, and " Insect Transformations," on the 15th, was followed by " Bees " next day, by " Ants " on the 17th, " Wasps " on the 18th, " Spiders" on the 20th, and " Reptiles " on the 21st, while " The Horse " was given twice on the 22nd, the first time at 4 p.m., and again at 7.30. Visitors from the neighbourhood were admitted as before, and all were delighted, each lecture proving more interesting than the last, while the liberal use of the coloured chalks added greatly to the effect of the sketches.

L

Next came a tour of some length, the towns visited including Sheffield, Maldon, Chelmsford, Lancaster, Birmingham, Doncaster, Leeds, Sheffield again, Ulverston, Birmingham again, and Leamington.

At the first of these places occurred the first of many more or less severe injuries to the frame. Throughout my father's career these accidents occurred at tolerably regular intervals. Perhaps the railway porters, in their eagerness to get the luggage out of the van with as little delay as possible, would tumble the great black parcel out unceremoniously upon the platform, from the height of a couple of feet or more ; and then an hour at least would have to be spent in repairing damages. Or a cabman would lift it to the ground with insufficient care ; or, more frequently still, willing but unnecessary and undesired helpers, with the best intentions possible, would carefully put together pieces which did not belong together, force the screws, wrench the ironwork, and so create havoc which it was oftentimes difficult to repair. Many were the makeshifts which my father was compelled to adopt at different times, and fortunate it was both for himself and his audiences that he was no mean amateur carpenter, and could often, by the aid of some ingenious temporary device, erect and use his screen, even after it had received some serious and unrepaired damage.

And by degrees, growing wise from experience, he greatly strengthened the frame. The ends of every piece of woodwork were carefully plated with iron ; the system of packing was improved ; the case itself was altered for the better. But even all these precautions failed to prevent an occasional accident, and always, at the end of a tour, the best part of a day had to be spent in careful examination of every part of the frame, and in various little repairs rendered necessary by the knocking about which it had undergone during its travels.

During the autumn followed lectures at the Sunday School Union, and also at Brighton, the Royal Naval School, New Cross, Norwood, Bedford, Streatham, Peterborough, Newcastle, Morpeth, Banbury, Windsor, Maidstone, Staleybridge, Saltaire, Huddersfield, and Liverpool (at the " Young Men's Christian Association "). This last institute comes in for almost the highest praise that my father ever bestowed upon any of the halls in which he lectured. I find the following notes in his diary :—

Capital hall at Y.M.C.A.; a good-natured hall-keeper, *who waits for orders.* Carpeted floor, and pitch pine ; *not object to screws.*

Then came the Christmas recess; but on January 3rd the lectures began again with " Life under Water " at Upper Norwood. At Sydenham, on the 8th and 15th, were given " Ant Life " and " Spider Life," for the benefit of various parochial institutions. Then came lectures at Caterham, Streatham (three delivered in the drawing-room of a private house), Romsey, Winchester, and the London Institution ; this last on a day of deep snow, when a cab could hardly be procured for love or money, and we were obliged to drive from the

L 2

Elephant and Castle station with two horses placed "tandem" wise. After this came the first of the Scotch tours, a lecture at Chester being delivered on the way, and succeeded by others at Stirling, Edinburgh (2), Falkirk, Dumbarton, Dumfries, Dunse, Kirkcudbright, Dollar, and Helensburgh. On the way home a stay of a few days was made at Conisborough, near Rotherham, and a short address delivered on "Ants," three or four spare leaves from a dining-room table serving as make-shift blackboards. Then followed engagements at Manchester, Stafford, Weymouth (2), Worcester, Harborne, Cardiff, Malvern, Norwood (5), and Yarlet Hall, near Stafford (3); and the season closed with "Insect Transformations" at Marlborough College on June 4th.

Seventy-four lectures only had been given, as against eighty-three in the preceding season, but as no less than twenty of the latter consisted merely of the short "gorilla" addresses at the Crystal Palace, which certainly were not "sketch-lectures," and perhaps ought hardly to be dignified with the title of lectures at all, the year's engagements were in reality rather more numerous than those of the preceding season.

In the course of his career my father naturally received many curious and amusing letters. He was accustomed to the correspondent who writes asking for an amount of information which, if given in full, would require about as much space as that occupied by an ordinary three-volume novel. He was familiar with the twelve or fourteen closely written, and perhaps crossed,

pages in which is embodied a minute description of
some perfectly well-known fact, which the writer nevertheless looks upon as a perfectly original discovery, of
a nature which will shake received science to its very
foundations. He was inured to the flat and sometimes
even angry contradictions of his statements which
periodically arrived, and which were not always couched
in the mildest and most inoffensive of language. As
for instance :—

> SIR,
>
> You say in your " Common Objects of the Country " (page
> 70) that insects never grow. You do not understand what you are
> writing about. Sir, did you never see black-beetles in your life ?
> As you seem not to have seen them, I send you a lot which I caught
> myself in my kitchen, and they are all of different sizes. Insects do
> grow, and you are quite wrong. I hope that you will publish this
> when you write again about insects.
>
> Yours indignantly,
>
> A. B.

But for cool audacity I do not think that he ever
received any communication to be compared with one
that reached him just about this time, and which
appeared to have been written in perfectly good faith.
The writer began by stating that my father had
deservedly risen to very high eminence as a naturalist
and an author, that his name had now been before the
public for many years, and that his reputation was
thoroughly established. He then proceeded to say
that, as one who had written and lectured so much must
have necessarily amassed an immense fortune, it was

clearly my father's duty to retire in favour of a younger man. And then he remarked that he himself was a younger man, and that, as he was very anxious to take up public lecturing as a profession, he would venture to ask my father to rest contented with what he had, done, and to transfer to himself " the goodwill of the business ! "

I do not know whether my father ever replied to this modest and unassuming epistle, but certainly he always treasured it as one of the greatest curiosities which he had ever received.

The 1881-82 season began rather later than usual with a lecture before the Gipsy Hill Band of Hope on September 19th. This was followed by " The Horse " at Tonbridge Grammar School ten days latet, and a somewhat prolonged tour followed immediately after, beginning with ": Jelly-fish " at Altrincham near Bowdon, on October 3rd, and ending with " Life under Water " at Brighton on the 27th of the same month. Between these dates fifteen lectures were given, with the same invariable success. Then came " Life under Water " at Newbury, which behaved rather better than on the occasion of his former visit ; then three lectures were given at Dover on successive evenings, while a second lecture at Brighton followed on November 5th, and a third a week later, the series being organised by the local Ladies' Committee of the Royal Society for the Prevention of Cruelty to Animals.

Travelling all day on the 13th (Sunday), Southport was reached late at night ; and on each afternoon of

the ensuing week a lecture was given in the Winter
Gardens. On the 21st the lecturer was at Leek, on the
22nd at Hanley, on the 23rd at Leighton Buzzard, on
the 24th at Coventry, and on the 25th at Worcester.
Then came two days of relaxation, and then on the
28th " Jelly-fish " was delivered at Weymouth. Next
day a second lecture, this time on " Bee Life " was given
to the Gipsy Hill Band of Hope; and on the 3rd of
the following month the first of many lectures was
given at Uppingham School, a place which my father
always greatly enjoyed visiting, and at which he was
especially popular. Three days later " The Whale "
was given at Stone, this time with a large drawing of
the animal, showing its internal anatomy in a peculiarly
striking manner ; and three days later still the ever-
popular " Unappreciated Insects "—now, however, considerably altered and improved—was delivered in the
Mechanics' Institute at Bolton.

Nothing of any special importance appears to have
occurred during this somewhat extended tour, and my
father's notes and ledger-entries are very scanty. At
Coventry, however, he seems to have met with a
novelty in the form of a feminine hall-keeper, ex-
pressively described in his note-book as " a starched
widow, who shakes her head and contemplates the
zenith."

On December 12th he was at the London Institution
for the third time, on this occasion with " The Hoof of
the Horse " for his subject. And he now succeeded in
obtaining an extra twenty minutes for his lecture, the

two previously given having been strictly limited to an hour apiece. Thence he went to Peterborough, to find, on arrival, that the hall-keeper, unauthorised, had let the hall for the purpose of an examination, and that it would not be free until the time came to open the doors for the admission of the public ; in other words, that the drawing-screen could not be put up. Consequence, a storm of indignation on my father's part, a threat there and then to leave the town without delivering his lecture at all, and the final submission of the keeper, who contrived to clear the hall by half-past six. Next day came " Unappreciated Insects " at Nottingham, in the Mechanics' Institute, and on the day after he was at Stamford, with the same subject. Thence he went to Driffield to deliver the same lecture again ; and so closed the first part of the season.

He was soon at work again, however, for December 30th saw him at Falmouth, for the first and only time. Then came a short recess, until Bowdon was visited on January 9th ; this for an " experimental " lecture, which was very well attended. The month was not a busy one, however, only six more lectures being given, namely, at Holloway, Salisbury, Slough, Redditch, and Hull (2).

February was better filled, with no less than sixteen lectures, including two at Clifton upon the same day (one at the College and the other at the Victoria Rooms), one at Cardiff, one at Weymouth, and three at Rugby. At Cardiff another feminine hall-keeper was met with, and was described in terms even less flattering than those applied to her predecessor at Coventry. "Sulky and peppery female," I read in the ledger; "regular Xantippe, so gave her no tip."

The three Rugby lectures were the first of a school course of six, the remaining three being given upon the first three Thursdays in March. In that month also came engagements at Yarlet Hall, near Stafford (2), Uppingham School (2), and Ascot. In April one regular lecture only was delivered, on the 27th, at Tonbridge Grammar School. Three engagements followed in May, at Yarlet Hall (2), and at Ascot again ; while June was more busy. For in this month, besides single engagements at Chislehurst, and one or two small schools, began a long course of lectures at the Crystal Palace, which were continued at intervals until the middle of September. Unfortunately, the hour selected (11.30 a.m.) was so early as to preclude visitors from a distance from attending these lectures, but the season-ticket holders and other inhabitants of the neighbourhood came· in considerable numbers, and the series was very successful. While this course was still in progress my father received the following letter from the then manager :—

CRYSTAL PALACE COMPANY,

CRYSTAL PALACE, S.E.

DEAR MR. WOOD,

I have much pleasure in stating that the Sketch-Lectures which you are now delivering at the Crystal Palace have attracted great attention, and seem to be as instructive as interesting.

I sincerely trust that we may be able to make arrangements by which you shall continue to give lectures here, as they are popular with all classes.

Yours faithfully,

(Signed)  S. FLOOD PAGE,<br>Manager.

Besides this course at the Crystal Palace, a series of lectures were also in progress at two schools at Upper Norwood, an arrangement being made by which my father delivered a lecture at each on every Saturday in term-time when he happened to be at home. And so, when, after a second course of six lectures at the Winter Gardens of Southport, exclusively upon marine subjects, the season came to an end, no less than one hundred and twenty-one lectures had been given in all, as against seventy-four in 1880-81, and eighty-three in 1879-80. This was the busiest year that my father ever had, as far as the number of lectures was concerned, although in several subsequent seasons the amount of travelling involved was considerably greater.

Yet it would be a great mistake to suppose that the profits, even of so successful a season, were very large. I do not think that, at any rate in this country, my father ever received more than ten guineas for a single lecture; and his average gross fee certainly did not exceed half that amount. Now when from five guineas has to be deducted the cost of perhaps one hundred and fifty miles' railway travelling, with a hundred-weight at least of excess luggage, the expense of cabs, tips to hall-keepers and porters, and the agent's fee, to say nothing

of the outlay upon chalks, printing, and postage, and
the wear and tear of the drawing-frame, it will easily
be seen that the net receipts are not so very high after
all. Many a time, especially in the case of a single
isolated lecture, has my father barely cleared his expenses; and I do not suppose that in any one year he
succeeded in clearing three hundred pounds by his
lectures. And so bad a man of business was he that I
am quite sure that, had it not been that all arrangements were in the hands of his agent, he would never
have contrived to realise even half that amount.

With one season prolonged until September 12th,
the interval between that and the next must naturally
be so brief as scarcely to be noticeable at all; and, in
point of fact, only eight days elapsed before the season
of 1882-83 began with the "Entomarchetype," or
"typical insect" lecture, at a large ladies' school in
Upper Norwood. Then, in the same month, followed
lectures at Chislehurst, Tonbridge, and at a boys' school
in Norwood; and with October work began in earnest.
During this month Maidstone, Upper Norwood, Chislehurst, Felstead Grammar School, the Royal Naval
School at New Cross, Leek, Rossall, Kilburn, and Yarlet
Hall, were all in turn visited, some of them two or even
three times. In November visits were paid to Cardiff,
Clevedon (where a comparatively scanty audience on the
first evening was replaced by one crowded even to overflowing on the second), Axminster, Upper Norwood (3),
South Norwood, Bedford, Hurstpierpoint, Warminster,
Chislehurst, and Coventry. And the year's work was

completed by the fulfilment of engagements at Hurst-pierpoint again, Reading, Upper Norwood, Hull, Nottingham, Stamford, Perry Bar, Driffield, Bradford (3), Redditch, and Ambleside; forty-two lectures in all during the first part of the season.

The second part was busier still, and the total number of lectures given very nearly equalled the record of the preceding season.

The first lecture given in the new year was also the first of a private series—six in all—delivered by special request at the house of a well-known London banker. They formed, in fact, the *pièces de résistance* of a course of social conversaziones, to which a large circle of friends were invited, and which were one and all thoroughly successful. All were given between the 3rd and the 15th of the month, at four o'clock in the afternoon; and universal regret was expressed when the course was concluded.

Still more interesting was the lecture given on January 5th, which initiated a series of twelve at the Royal Normal College for the Blind, at Upper Norwood. It seemed rather a contradictory arrangement—*sketch*-lectures to those who could not see. But the sketches were nevertheless executed just as usual, and were explained to the patients by the attendants; so that, although unable to see the drawings for themselves, they were yet enabled to form a tolerably clear idea of what those drawings were meant to represent. And they were all thoroughly delighted, both with this and with the subsequent lectures of the course. They applauded the lecturer to the echo; they were most eager with their questions as he was leaving the platform. And no doubt they gained almost, if not quite as much, amusement and instruction as if they had possessed the full use of their eyes.

The other lectures given during the month were at Bristol, Weston-super-Mare (2), Highgate (2), and the Birkbeck Institution; and these were followed in February by others at Tonbridge Grammar School, Brighton (3, a marine course), Leamington, Harborne, and two of the Upper Norwood Schools.

March was a busy month, as usual, with seventeen lectures, including two at Uppingham School, one at Marlborough College, and others at Worcester, Tottenham, Tonbridge Wells, Tonbridge, and Lytham. April brought one engagement only, at Malvern, besides a couple of the occasional Saturday school-lectures at Norwood. May brought seven—two at Chislehurst, three at Yarlet Hall, and two at a private school at Earl's Court; June, six, including three more at Chislehurst, two at Earl's Court, and one at Forest School; and July, three, which brought the season to an end. The total number of lectures given amounted to one hundred and thirteen.

A great change was to take place in the following season. My father had, several months previously, received a pressing invitation to cross the Atlantic, and deliver the "Lowell" Lectures for the autumn of 1883, at Boston, Mass., U.S.A. For some time he had hesitated; partly by reason of family ties, partly from the distance to be travelled, and partly from a doubt as to whether the English connection might not suffer by his temporary withdrawal from the scene of action. At length, however, he arrived at the conclusion (a false one, as events afterwards proved), that his absence for a winter would lead to an increase of the number of lectures in the following season; that an American tour was likely to prove profitable; and that in other ways, besides that of mere lecturing, he might make the trip a successful one. And so his decision was made, and the American offer accepted.

As, however, he was bound by the terms of his agreement not to lecture in America before giving his course at the Lowell Institute, he thought it advisable to begin the season in England, and not to cross until he could delay no longer; there being, of course, no object in his staying idle in America while he might have been earning money in England. Therefore, lectures were arranged for and given on September 28th, at the Royal Naval School, New Cross; on October 1st at Coventry (a special lecture on the Horse); at New Swindon two days later; at Yarlet Hall on the 8th and 9th; at Wolverhampton on the 11th; at Westbourne Park on the 12th; and, finally, at Sheffield on the 15th.

Meanwhile, preparations for the transatlantic tour had been going on apace. Two drawing-frames were thoroughly overhauled, put in perfect repair, and fitted with new canvas. Notes had to be drawn up for the special series at the Lowell Institute, on which, probably, the success of the trip would depend, while all the necessary preparations for a six months' absence from England had to be made; all this in the occasional flying visits home, which were all that could be managed in the three or four preceding weeks. But all was at last completed, and on October 15th my father left home, not to return until well on in the following spring. The same evening he gave his Sheffield lecture, on Ants, and on the next day reached Liverpool, to embark at once upon the Cunard steamship *Cephalonia*. On the following day the vessel left the Mersey, touching at Queenstown a few hours later, to pick up passengers and mails; and by the morning of the 19th she was well out at sea.

# CHAPTER XII.

## THE FIRST AMERICAN TOUR.

BEFORE setting out upon his first Transatlantic tour, my
father made an arrangement with us at home, in virtue
of which he was to write letters whenever possible, but
was in addition to keep a daily "log," portions of which
were to be sent off whenever an opportunity should
occur. This undertaking he faithfully carried out, and
almost regularly once a week, and sometimes more

frequently still, a bulky letter used to arrive bearing an American postmark, and containing eight, ten, or twelve closely-written pages of diary. This, when it arrived, was copied into a large manuscript book, and thus we have, in his own words, an accurate record of almost the whole of his doings during this his first visit to Transatlantic shores.

The passage from Liverpool was a very rough one.

I knew that the storm-cone meant a lot (he says), and was not mistaken. We *are* just catching it, and no mistake ! Crash ! A shower of broken glass has just come into the saloon, having been pitched out of the "safety" trays. I have been obliged to wedge up the inkstand with books, &c., in one of the table trays, as it kept sliding about the table in the most absurd way. There is not much pitching, but the vessel is rolling greatly, and when a big thing like this takes to rolling it "goes the whole hog." Which word makes me think of rats. At breakfast this morning I heard one of the passengers tell another of his dog, who caught a rat " by his pants," just as he was diving into his hole. I thought it a beautiful euphuism . . . At present—10.15—there is only one passenger in the saloon except myself. I would go to bed myself, only there is little use in it when fellow-creatures are afflicted. In room 62 there are two ladies and their maid, and which is the worst I don't know. I think, however, that it is the maid. As there is only a plank between the rooms, every sound is audible from one room to the other.

October 19th, 8.45 p.m. *Such* a night, and no wonder that we saw a Mother Carey's chicken this morning. My portmanteau got loose, and danced and jumped all over the floor as if it were a live thing ; and how I escaped being pitched out of the bunk is a wonder. This morning the few passengers who appeared at the breakfast-table were comparing notes as to knees and elbows. Unwittingly I deprived one of them of his breakfast. Being in a state of outrageous hunger, I had ordered liver and bacon for breakfast. My opposite neighbour, who has crossed several times, was just making

M

his way very slowly through a cup of coffee when my liver and bacon were set before me. He looked at them for a moment, turned pale, and fled. Just as I went on deck, a message was sent to me that the chief steward had ordered a plate of grapes into my cabin. So I distributed most of them among the sick ladies, took a few myself for form's sake, and gave the rest to my victim.

Early this morning I went on deck, and was enchanted. First I went to the extreme stern. Up it went, and then down, down, thirty feet down. Then it paused for a moment, and from the screw there rushed a torrent of liquid sapphire edged with emerald, and extending as far as the eye could reach. Then to the bow, which looked like a gigantic plough, flinging on either side vast swathes of sapphire, flecked with foam as white as snow. As the ship passed on, one could see the sapphire masses sinking deeply into the black depths below. Then the wind tore the edges of the swathes into clouds of spray, in which a succession of rainbows played. I calculated that on an average each wave raised my bed at least as high as our house, and, of course, dropped as much, while the roll was about fifteen feet. The officers and stewards are quite astonished at me, for I feel as lively as a dozen crickets, and resemble Dr. Gordon Stables' cockroaches in point of appetite. One of the passengers had put his portmanteau on the sofa when he went to bed. In the night the ship gave a tremendous roll, and the portmanteau was pitched completely across the cabin, falling on his stomach. The captain never went to bed after we started until one this morning, when he turned in for a short time. By the way, we saw two great schools of porpoises, many of which came quite close to us, showing the wonderful grace of their leaps. Despite our speed they caught us with the greatest ease, kept pace with us for some time, and then crossed our bows as if we were standing still. The deck is such a curious sight when the weather is fine. It is covered with rows of passengers in deck chairs, some sitting, but most reclining. Passengers have to supply their own chairs. Mine is a very modest business, but most of them are couches, and would make very fair beds. As to the ladies, they might be Egyptian mummies, for they are tucked up in rugs and furs up to their throats, and have shawls wrapped round their heads, so that the occasional tip of a nose is the sole indication of a human countenance.

October 20th.—Wind still fiercer. At five o'clock this morning there was a slight lull, and I was nicely asleep, when the men must needs holystone the deck just over my head. The noise beats the shunting all to fits.

We were living near a railway terminus just at this time, and the goods traffic was carried on principally during the small hours of the morning: hence the allusion.

Holystones look like gigantic Bath bricks fixed at the ends of broom-handles. First, the deck is sluiced, and while it is wet brick powder is scattered over it, and two men run the holystone up and down, making a noise much like the escape steam of a locomotive. This went on until seven, when I had had enough of it, and got up. I mean to turn in early to-night, for the noises on board a steamer are much the same at all hours, and nothing is to be gained by waiting.

This morning the waves were so huge that, when I stood at the stern, they rose above my head like mountains of blue glass. The wind veered northward about mid-day, so that some sails were got up, and we are much steadier. Since noon yesterday we have run 234 miles, and hope to do better by the same time to-morrow, as the wind is not directly ahead. Please remember that the vessel is pitching some thirty feet, rolling about ten feet, and throbbing all over with the engines, so that my writing can't be expected to be very good. Nothing eventful occurred to-day, except that we shipped a sea which swept over the upper deck, evoking great screams from the emigrants on the lower deck, and the ladies on the upper deck. I was at dinner at the time, and saw the great wave climb over all the ports in the most leisurely manner.

The captain asked if I would preach to-morrow morning. So I said that I would give an unconventional discourse, and pleased him much by offering to do the same in the evening.

October 21st.—Everyone delighted with morning and evening, and requests all round for repetition. This morning the wind lulled partially. One of the stewardesses told me that she had not been to bed since we started, *i.e.*, for four consecutive days. One lady has

not yet left her berth at all. To-day, however, several new faces
appeared at luncheon, and it is actually possible to take a meal without clutching frantically at the plate to prevent it from jumping into
one's lap. I think it immense fun, but stand alone in that opinion.
By the way, everyone *will* call me "Doctor." It is of no use explaining. They all say that they have been accustomed to call me
"Doctor" as long as they can remember, and think they cannot call
me by any other title.

Throughout my whole memory of him my father
was always very contemptuous with regard to the
"Mizpah" rings which have been and still are so
fashionable; and he never lost an opportunity, in sermon or in ordinary conversation, of inveighing against
them, and showing the fallacy of the misinterpretation
to which they owe their existence. For, as he was continually pointing out, the sentence, "The Lord watch
between me and thee when we are absent one from
another" (Gen. xxxi. 49), was not spoken in a friendly
manner at all, but merely implied that neither of the
covenanting parties could put the least faith in the
other. So far from being a prayer or a blessing, in
fact, it was practically an implied curse. And he seems
to have taken the opportunity of working this into one
of the "unconventional discourses," for on October
23rd, the "log" runs:—

It is so absurd. Lots of the passengers came up to thank me for
my semi-sermons, and Judge F—— was very emphatic. He said
that they never got such sermons in America, doctrine and florid
eloquence being paramount. He was delighted with "Mizpah," and
ever so many of the passengers came to me lamenting that they had
sent Mizpah rings, &c., to their friends before starting. This evening I gave the "Cockroach," putting up with a tarpaulin very

indifferently stretched.  However, the lecture was an immense
success.   .   .   .

October 23rd.—Wind very fierce again, and is now shrieking
through the rigging.  Just as I came on deck this morning a
man fell off the main-yard and was killed on the spot.  We left
him in mid-Atlantic at four p.m.  Of course the event threw a great
gloom over the ship.  I have been asked on all sides to give
another lecture, and if the sea behaves itself shall do so to-
morrow.  Everyone is talking about the cockroach, and some of
the passengers have been cockroach-hunting in the surgery with
the aid of an electric lamp about as big as a marble.   .   .   .

Bless these fragile American ladies !  How they do eat !  They
have had three huge "square" meals, have been nibbling gingerbread
and eating fruit all day, and now (10 p.m.) are going in for supper.

Next follow some minute directions for getting into
one's berth during bad weather, which I quote for the
benefit of inexperienced travellers.

Getting into bed is a Science.  First you roll up the clothes as
small as possible against the wall.  Then you hold on tight to the
edge of the bunk, wait for the roll of the ship, and get the right leg
over.  On the next roll you only have to jump, and are chucked into
the berth.  Then you "scrouge" yourself against the board, and by
dint of wriggling and coaxing you get the clothes over you.  Once in
you *must* lie on your back, and quite straight.  If you try to bend
either legs or arms your knees and elbows are covered with bruises.
I always "play at" being an Egyptian mummy, and somehow can
sleep better than at home.  Only it is necessary to pad yourself
tightly against both sides, as otherwise you will roll against each side
alternately as the vessel rolls.

There seems to have been but little fair weather
during the passage, which must have severely tried even
the most experienced sailors.  On the 24th of October
is the following entry :—

This morning I submitted to the captain that the Atlantic was

estimable in its place, but that it was out of place in my room. I
hope never to have such another night. The wind was terrific, and
more than a storm. Pressure on the wind-gauge of twelve degrees
means a hurricane, and we had eleven degrees, and right in our
teeth ! China and glass were crashing all over the ship, and at 1.30
there came a terrific bang, followed by shrieks, and accompanied by
a torrent of water rushing through my room. Then the vessel went
up somewhere, and back came the water. Then there was a great
turmoil on deck, shouting and stamping of sailors, screaming of
women and frightened children, &c. Just to show what a storm it
was, I found an empty drawer from a room at some distance lying at
my door. It was all I could do to keep in my berth, and one lady
was pitched completely out of hers into several inches of water. It
was so bad that two boats were slung ready for immediate launching,
and a life-belt was laid out for each passenger. You may imagine
what a noise the Irish emigrants made. Knowing that I should
only be in the way if I got up, I just held on tightly. There was
not much sleep for anyone that night. At daylight I got up and
found that the portmanteau had kept out the water, so that, except
soaking boots and slippers, no harm was done. Some of the
passengers, however, had nearly all their clothing spoiled, and one of
the officers was washed out of his berth. We had shipped a sea
astern, and it had broken into the main-deck, sweeping it fore and
aft as the vessel pitched. Yet I had not the least indication of sickness, though scarcely another passenger escaped. There was one
exception, a young lady, but old sailor. She foresaw what was
coming, and did not undress. About noon, the wind went round to
the north and the waves abated, but the long Atlantic swell continued until evening. Then, at the request of the passengers, I gave
another lecture. . . .

Here we are, a week out, and barely half-way to Boston. We
only ran 213 miles in the last twenty-four hours—an easy six hours'
journey by rail. It will be a very close shave for me to arrive in
time. The captain says that he has hardly ever been as late as
Sunday, but that this time he doubts whether we shall arrive until
Tuesday morning. Just fancy what it must be to those poor
passengers who are prostrate in their berths, and have had a week of
sickness. The ladies in the next cabin were too ill even to be
frightened.

October 25th.—Just off Newfoundland Bank. A whole flock of various gulls are coming quite close to the ship. Yesterday, or rather the day before, we saw many Mother Carey's chickens. Their flight is very curious. They keep close to the surface, wheeling up and down the huge waves in a manner resembling a mixture of the bat and the swallow. This morning was wonderfully quiet, so that the guards were removed from the tables. Now, however, the vessel is rolling as badly as ever. About 5 p.m. we got a taste of the Newfoundland fog, and the horrid fog-horn began to howl. During the last twenty-four hours we ran 283 miles, so that there is, at all events, an improvement. Yesterday the vessel gave a tremendous lurch, and four ladies shot out of their chairs and slid down the deck until brought up by the bulwarks. One of them, who was eating an apple, stuck by her plate manfully. I congratulated her on her courage, but she said that it was so sudden that she thought herself still seated in her chair, and that she clutched her plate because there was nothing else to hold by. . . .

They have such a queer custom. The dessert is laid out on the tables, and as soon as they sit down the Americans lay violent hands on it, so that before you can look round scarcely a pear, apple, orange, or bunch of grapes is to be seen.

After this date, the weather seems to have changed for the better, so that the dismal prognostications as to delay in arriving were not realised. On the evening of the 26th advantage was taken of the comparative stillness, and another lecture was delivered—the second part of "Pond and Stream," which had been begun two days earlier. The unconventional discourses of the preceding Sunday still seem to have afforded a leading topic of conversation :—

Mr. P—— (nominated as Governor of one of the States) paid me a very pretty compliment. There was a conversation about various preachers, and Mr. P—— turned suddenly on me, and " guessed " that no one went out when I preached. The sermons

seem to have made an extraordinary sensation, and judges and
senators are continually asking me to explain passages which they
could not understand.  At last, I was obliged to say that I really
could not be expected to answer theological questions generally.

October 27th.—Curious variations of temperature.  We got out
of the Gulf Stream and were very cold.  Then we got into an elbow
of the stream and were warm.  Now we are in the north-east
current and are again cold.  The last two days have pulled us up
wonderfully, and if there be no fog we ought to sight land by
noon to-morrow. . . .  I have found out the mystery of the dessert.
It is all pocketed to be eaten before breakfast next morning.  They
are always at it, and it is no wonder that half the passengers have
artificial teeth.  Toast and jam, "candy," apples, oranges, celery,
and buttered toast go on all day, accompanied by uncounted quarts
of iced water.  The consumption of ice is almost incredible, and I
almost wonder that they do not pocket the ice, as well as the fruit,
biscuits, and "ginger-snaps," &c.

On the 28th, land was duly sighted, after—in these
days of ocean racing—a prolonged passage of nearly
eleven days.  My father had contrived during the
voyage to make great friends with a fellow passenger
of influence, who undertook to see him safely through
the Custom-house.  But when the time came he
found that mediation was vain, and that he must
patiently await his turn with the other passengers.  He
employed his time in "taking stock" of the building,
and does not seem to have been greatly struck by what
he saw there:—

Oh, that Custom-House! (he says).  Imagine a vast wooden
shed, very dark, and smelling like a stable.  Truck after truck full
of luggage dumped down anywhere, and the passengers left to pick
out their scattered property as best they can.  If you appeal to an
official there is only one answer : "Git all your own things together
and then wait till your number is called."  One lady had nine

" Saratoga " trunks, each big enough to receive a dress without folding, besides boxes, bags, parcels, deck-chairs, &c. I had trouble enough myself, although an old traveller, and when I saw one of my belongings in a truck I stuck to it, and would insist on the men putting it in my corner. I should think that at least two acres of ground was covered with luggage. Then, large females *would* sit on boxes which did not belong to them. Fortunately, Mr. P——, who is a very great man here, introduced me to the chief officer. So neither the frames nor the bundle were opened, and the trunk, portmanteau, and bag were only opened and shut again. . . . As to Boston, except the very modern part, it is as dull, dismal, and squalid as any part of Deptford or Wapping, and if a square half mile of each were transposed, I do not think that any·one would see much difference. The chief distinction is that the inhabitants live in flats, and that the houses are red brick, with outside green louvre blinds to each window.

Next day, the 30th, the preparations began for the first " Lowell " lecture, which was to be delivered on the day following. Judging by the entry in the " log " for that day, American hall-keepers are very much what they are in England.

Mr. Lowell called this morning, accompanied by Dr. C—— , the manager. We went over together and sent for the frame, which was unpacked, and I told the keeper how to lay out the pieces, so that after breakfast I could put them together. About 11.0 Dr. C—— came again, and said that the janitor had put up the frame, but did not quite know how to put up the canvas! I found that he *had* put up the frame! All the screws were outside, and every bit was wrong. He thought that it would save time. So it had to come to pieces again, and the janitor said that he felt "real mean." However, he is full of admiration now that he sees how it is managed. . . . .
October 31st.—Yesterday was taken up with preparations, so that I could not write. There was an unexpected run on the tickets, so that the applicants had to be formed in *queue*, two and two, and get their tickets in their order. Dr. C—— said that it

was the largest house that he had known for thirty years. After the lecture, he said that it was the very thing which was wanted. The papers are so full of the elections that there is no mention of the lectures, though there may be this evening. Even Irving only has a few lines, and so has Matthew Arnold.

November 1st.—Just had a long interview with C. H. H——, of the American agency—satisfactory as far as preliminaries go. I gave him nearly all the copies of the syllabus. His brother-in-law, a scientific party, was at the Lowell Institute yesterday, and spoke in the highest terms of the lecture. It appears that I am legally entitled to the prefix of "Professor." However playfully conversational titles are given, ranging from General, Colonel, Judge, Doctor, and the like, they are never used in print or writing unless they be genuine. Mr. Williams was very strong on the subject, and insisted on substituting "Professor" for "Rev." He says that the clergy are, as a rule, so conventional, that people are shy of going to hear a "Rev." lecturer. . . .

How would you like this kind of luncheon? Clam chowder, Mayonnaise chicken, corn-cake, ice (Ponche Romaine), grapes *ad libitum*, and coffee. If you like you can begin to breakfast at 7 a.m., and go on until luncheon. Then you can eat luncheon until tea-time, and then dine until supper time, when you can go on with supper until 11.0. As to fruit, whole piles of apples, pears, and grapes are on every table at every meal. I always pitch into the grapes. There is a black grape here quite new to me. It is not very large, and has something of the Muscatel flavour, but is sweeter. It is called the "Martha." The peculiarity is that as soon as it is bitten the whole interior tumbles out of the skin. . . . .

Clam chowder is something like green-pea soup, with dried oysters grated into it. Alone, the clam is rather too strong. Fried oysters are stunning, each being separate and bread-crumbed. . . .

The Archæological Institute of America has sent me an invitation to a conversazione this evening, and to-morrow I am going to the meeting of "Scientists," as men of science are called here.

This morning I went to explore. Tremont Street is wonderful. On one side are the best shops, on the other the gardens, and in the immensely wide road is a four-fold stream of tram-cars ; " horse-cars " as they are called here. Imagine Regent Street twice as wide, and

having Hyde Park on one side. In almost every tree there are
"martin-houses," *i.e.*, Swiss cottages, made of wood, for the benefit
of small birds. The police are dressed in blue-grey, with grey
helmets. They do not have a sheath for the baton, or "club" as
they call it here, but carry it hanging by a string from the hand.
It is small, but heavily loaded.

On November 17th, however, there is a second note
upon this subject:—

The beings whom I took for police are park-keepers. The real
policeman is a big man in dark blue, with a helmet like that of our
own "Bobby," and a staff which he carries in a ring (not a sheath)
attached to his belt. Also, he has a silver shield on his manly
breast, and he has a revolver in his pocket. On the whole, however,
he is fairly affable, and though he *is* said to have his helmet full of
cigars given to him by provident bar-keepers, much the same is said
of our own specimens.

A further note, however, dated January 7th, 1884,
shows these officials in a somewhat less favourable light,
and also shows something of the practical results of
compulsory Sunday closing:—

You will see in "Puck," which I am sending you, a mention of
police who go over to Hoboken on Sunday to get a drink. Each
State makes its own laws. If Massachusetts were to make a law
that everyone *must* carry a loaded revolver and smoke six cigars
daily, the Government at Washington could not interfere. In New
York, at present, there is a law which closes all bars on Sunday, and
prohibits the sale of wine, beer, and spirits. Consequently, there is
more drinking on Sunday than on the week-days, only it is done on
the sly. The known frequenters of hotels go boldly in at the front
entrance, and so pass into the bar, the blinds of which are virtuously
drawn down. Others not connected with hotels have private signals
with their customers. The man walks past the door and gives the
signal. Then he saunters back again, leans against the door, which

opens for a moment, and in he goes. Oddly enough, the policeman always happens to be looking in the opposite direction, so that Sunday is a profitable day with the force. The police, however, dare not go into a bar themselves. So, when relieved, they go in the ferry-boat across the Hudson into New Jersey, where the prohibition law is not in force.

One very neat dodge was discovered lately. A portion of a brick wall was taken out, and replaced by a wooden door painted like bricks. It was not adjoining any bar-room, so that it was not suspected. But it opened into a passage which gave access to a bar on the other side of the block. Everyone knows that prohibition is a farce, and yet the officials are forced to make the law lest they should lose the vote of the teetotallers, who are as noisy here as in England.

Returning to the above-mentioned meeting of " Scientists," he goes on to say :—

In the evening I went to the meeting ; very interesting, but all the describers were wretched speakers. The only good speaker was the Rev. Phillips Brooks, of Trinity Church ; which reminds me that the American Prayer Book is almost entirely the same as ours. The Marriage Service is cut down quite three-fourths, and the Office for the Visitation of the Sick is also mutilated. The Consecration Prayer is essentially the same. Considering the strong Puritanical influences of the place, the book is most praiseworthy. . . . . Washington Street is all shops, and is three miles long. Most of the shops are upstairs. Locomotion here is almost entirely managed by " horse-cars." Cabs like ours do not exist. There are two-horse carriages called " hacks," but their charges are exorbitant, *i.e.*, two shillings for the shortest distance *for each person*, so that four people would have to pay eight shillings. Hence the horse-cars. Luggage is conveyed by carriers (" express-men," they call themselves).

They have the oddest names here. Plumbers are called " sanitary engineers " ; and a lift is an " elevator". . . . . The railway station is called the depot—pronounced " deep-oh," and you can buy railway tickets almost anywhere except at the station, and if you don't want to travel you can use the ticket as cash. It looks odd to see a man pay with a " five spot " (*i.e.*, five-dollar) note, and take as part of his change a railway ticket for a place which he does not mean to visit.

The " American drinks," of which so much has been said, are really very harmless affairs. There is barely a tablespoonful of spirit. Then there are several flavourings, then pounded ice, then it is shaken violently, then iced water and pounded sugar, then more shake, and then a shell-shaped silver colander is put over the mouth of the glass, and the mixture strained through into a tumbler. " Cocktails," I am informed, are taken *before* breakfast, and never afterwards. If you take a " straight," it is another business. The man hands you the bottle, a glass, and iced water. You may fill the glass with the spirit, if you choose, or only wet the bottom. This last is the plan usually adopted, where a refusal, no matter how well grounded, is taken as an insult.

There is a new fashion for ladies' dress. Instead of leaving a little bit of white handkerchief sticking out of the bosom, they put a stuffed bird spread out flat, the colours being due to aniline dyes. If they have no bird, a bouquet (pronounced " boh-kee ") takes its place.

The Lowell lectures were delivered bi-weekly, on Tuesdays and Fridays, and as, by the terms of the agreement, my father was precluded from lecturing elsewhere until the series (of 12) was concluded, he had during the earlier part of his visit a good deal of spare time on his hands. The second lecture was delivered on November 2nd, and seems to have been a great success.

Such an audience (runs the " log "), and such a success. Some six hundred of them were grey, white, or bald, with beards to match, and I heard that almost every man of science was there. Mr. Lowell was present, and I have established my reputation as originating a new epoch in lecturing. Mr. Lowell told me that the audience generally thought that I was the son of the author, and happened to bear the same name. People seem wild to procure tickets, and if the hall had been twice as large it could have been filled. Mr. Lowell says that I now may accept as many engagements as I like in or out of Boston, provided that I give no lecture until after it has been delivered at the Institute. The *New York Herald* announces my coming, and quotes the Boston papers. People here are curiously impressed with my pension. They would not think half so much of a C.B. By the way, instead of illustrating the development of an insect by the gnat, I took the *Eristalis* (drone-fly), and produced a perfect storm of applause. The iridescent wings took them all by surprise, and I heard them saying that nothing had been seen to compare with it. . . .

It is a good thing that I brought supplies with me, as nearly everything has been doubled in price since the war. Why clothes should be so dear I cannot imagine. A coat, for example, which would cost about £2 in London, costs from £6 to £7 in America. Consequently the knowing ones get themselves carefully measured at home, send the measures to London, and have the clothes sent to America. They have to pay 50 or 60 per cent. on the value—say £1 10s. at the Custom-house—but even with that and the freight they save at least £3 on the suit. American tailors are grumbling terribly, and are trying to put it down, just as our grocers, &c., try to put down the stores. . . .

Art is charged higher than anything else. For example, I bought one of our penny paint-brushes, and had to pay sixpence for it. Had it not been for Mr. P——, I might not have got my pastils out of the Custom-house yet. The officers have a sweet way of impounding things which they well enough know to be free, charging anything that they like, and then saying that you can pay *under protest*, and so have the right of appeal! The Court of Appeal is much like our Chancery, and if you win your case it half ruins you, and does so entirely if you lose it. One of our best artists—I think it was Herkomer—took with him some of his own pictures in order to illustrate his lecture. Previous to sailing he insured them at their full value. The Custom-house heard of it, and assessed the paintings at 90 per cent. of their value. He was so angry that he refused to clear them, and left them there until he returned to England. So the Customs got nothing.

As to variability of climate,· England cannot hold a candle to

Boston. This morning was so cold that some courage was required for a bath. Now it is so warm that I have been obliged to put on my light coat.

Another Americanism. In every room of the hotel, in every corner and gallery of the museums, there are " cuspadors." This is Yankee for spittoon. Mine is very handy, as I use it for a waste-paper basket.

Many of the Boston churches were visited in turn, and the "log" contains descriptions of several. Among the special points that seem to have struck my father, besides the use of the English Prayer-book, with the modifications already noticed in the Occasional Offices, was the distribution of functions when more than two clergy were present. The Roman Catholic Cathedral does not seem to have impressed him particularly, save as regards the excellence of the music. He gives an amusing description of a sermon by an Irish American :—

November is the "Month of the Dead," set aside for the purpose of freeing the souls of the faithful from purgatory, they, as H—— says, " being incapable of prayer for themselves." And the preacher gave us Purgatóry, as he called it, hot and strong. After saying that the pains of purgatory were the same as those of hell, except that hope was not taken away, he proceeded to describe the scourge, and the rack, and the wheel, and the stake, and said that all these agonies put together would not equal an hour of purgatory. Then he showed us how the faithful departed behaved in purgatory. He shrieked, he sobbed, he groaned, he cried for mercy, and nearly drove the females into hysterics by telling them to think of their loved ones departed as being thus in unspeakable tortures. The upshot, of course, being that the more money they paid for masses, the sooner would their friends escape. Otherwise, when they too entered purgatory, they would be met by the reproaches of their friends for leaving them still in that place of torment.

On November 5th torchlight processions perambulated the town, not in honour of Guy Fawkes' fiasco,
as my father at first supposed, but for electioneering
purposes.   Later in the evening a fire occurred in the
neighbourhood.   This resulted in a good deal of noise:—

The engine continually rings a big bell, which gives right of
passage.   The men yell and shout, and all the church bells of the
district bang at random.   The track of the engine is marked as far
as you can see by a train of burning coal ; which word reminds me
that this is a smokeless city, like Paris.   No one would think of
burning any coal except anthracite.   I have now learnt the meaning
of the word " build " a fire.   The coal must be broken into knobs
not larger than the asbestos balls in a gas-stove, and they must be
*built* in the same manner.   It takes a little more trouble at first,,but
there is none afterwards.   No smoky chimneys ; no soot ; no poking
or raking.   Only, when fresh coal is put on, it must not be shovelled
on anyhow, but laid on so as to allow air to pass freely.   The heat is
bright and clear, and there is no dust flying about.   For a kitchen
stove it would be invaluable, as the flues would never get choked.
. . .   " Coat " and " boat " are here dissyllables.   Tremont (three
mountains)· is   pronounced "Tremmont" ; Boston   is   pronounced
" Borston."

During the first week of the Boston visit the city
was in the full flow of electioneering excitement ; and,
seeing how absorbed all Americans become in political
affairs at such times, it is really astonishing that the
rate of attendance at the lectures was maintained.   Yet
the hall was crowded to overflowing upon each occasion
My father saw something of the manner of voting, and
has thus described his experiences :—

November 11th.—On Tuesday, Dr. C—— drove me to four
of the voting places, and I went through all the proceedings except

that of actually putting the ticket into the box. I had a little mild chaff with the officials, and suggested that if I voted impartially for all the candidates there could be no harm in their acceptance of my votes. I will send you a collection of tickets, which were quite affectionately pressed upon me. When you vote, you push your ticket endwise into a slit. There it is seized between two rollers, which stamp it and record the number on a set of figures under glass. Thus the number of votes which have been given is recorded, but not the proportion given to any candidate. The excitement has been tremendous, and in several places pistols were used freely, with loss of life and many wounds. Contrary to Mr. Lowell's expectations, there was no diminution in the attendance at the lecture. Yet in many parts of the town the streets were impassable, the horse-cars taken off, and all wheeled vehicles sent round by other routes. In several places the votes were shown on house fronts by means of the magic lantern, and as the vote of each district was exhibited, the yells were terrific. Here they do not cheer, but bark, with a series of short yaps.

The description of the American girls does not sound altogether attractive to English ears :—

If English girls used such astounding expressions as these girls do, they would be thought horribly vulgar. But, somehow or other, they "guess," and "conclude," and "allow," &c. &c. ; they ask the waiter for a "swallow" of water ; they use the word "fix" for everything except its right meaning, and then they say "settle." They talk of "running" a house, *i.e.*, keeping it. And yet, though they use their noses as organs of speech, they do not seem vulgar. . . . . I heard an eloquent preacher, by the way, ask in the pulpit for liberal contributions, because it cost so much to "run" so large a parish. . . .

What would you think of having spittoons—I beg pardon, cuspadors—in the drawing-room ? Here they are, but of elegant design—pink and gold china, about two feet high. Even in the elevator there is one, but only plain blue and white. *The height is to make them more accessible to ladies*, gentlemen being supposed capable of hitting the ordinary article at five or six feet.

N

Then there is a description of the American weather. My father, not quite knowing what to expect, made no special preparation for the winter before leaving England, save that he purchased a tolerably large stock of ordinary winter clothing. But he very soon discovered that for all practical purposes this would be well-nigh useless. It might do very well in what is generally considered as a " sharp frost," but for a climate in which the thermometer is apt to fall occasionally to some thirty degrees below zero, and in which the wind is given to blow with almost all the force of a hurricane, it soon proved altogether insufficient. So it was necessary to do in America as the Americans did, and purchase clothing better adapted to the vagaries of the thermometer.

November 12th.—We have had a fine specimen of Boston weather. Last night was so warm that I had to take off all the bedclothes. This morning, about 3.0, I awoke half-frozen. A fierce north-westerly gale was blowing, and literally howling round my room, which exactly faces that quarter. So I had to get up, fasten the window, replace all the clothes, and add the fur rug, and even then could hardly get warm. During the day there were several fierce snow and hail storms. In returning from the S——'s I half thought that both ears would have been frost-bitten, and regretted that I had not put on the travelling cap. . . But, for compensation, *such* a sky. I certainly never saw the moon before. She was so brilliant that it was barely possible to look at her without being dazzled, while Jupiter blazed with a splendour which he never attains in England. Few stars could be seen, the moon killing them all, and even Sirius and Capella were barely traceable.

On the 1st of December there is a somewhat similar note, to which is appended a further eulogy of the anthracite, or smokeless coal. I may add, by the way, that one of my father's first proceedings after his return home was to order a quantity of this coal for his own study fire; that he failed to make it burn; and that his enthusiasm in its favour thereafter considerably waned. Here is the note :—

Yesterday the thermometer, with its usual playfulness, dropped from 70° to 8°, and the Bostonian clerk of the weather again turned on the north-west wind; which, as William Black remarks, is enough to blow the ears off a brass monkey. So as I had much writing to do, and it was scarcely possible to hold a pen, I ordered a fire. It was lighted at 10 a.m., and at 1.30 was made up. Since that time it has not been touched, and now, to-day, at 11.20 a.m., it is still alight. What coal is there which would remain alight for twenty-two hours and then be too hot to be approached? There is no smoke, and only a solid mass of fire, with pale flames flickering over it. If I can have my way I'll save half our coal-bill. I know that nothing has been added for twenty-two hours, for, as I went out, I put a damaged tie on the fire, and there its ashes are still. There is only one defect (if it be a defect), and that is that the anthracite fire is as tenacious of life as a large-bodied moth.

A few days later there is another weather note :—

O! this weather! Here's a specimen. Saturday night thermometer at 12°. Sunday morning, at 6.30, snow and fierce wind. We go to church, and on coming out find heavy rain, and thermometer at 65°. Monday morning, thermometer at 8°, and skating on the Frog Pond. Tuesday, rain again, then blue sky and brilliant sun, and such a wind that the mud suddenly became dust. Yesterday (Wednesday) rain again and weather close, and too warm for an overcoat. At 2.0 p.m., snow again and cold wind. This morning, an azure sky, blazing sun, and north-easterly wind. No mortal can avoid taking cold. Even Americans from New York, a much colder place, cannot stand it, and I am told that, for a foreigner, I am enduring it wonderfully. . . . The Lowell Institute is in the very focus of the winds, Boylston Street debouching on a branch

N 2

of the Bay.  So the wind tears down the street, and anyone on the
top of the steps is exposed to its full violence.

And again :—

December 16th, 6.30 a.m.—Yesterday was another "brass-
monkey" day, the thermometer down to anything you like, a bright
blue sky, dazzling sun, and a north-westerly wind that seemed to
pierce any clothing except fur.  I had read of this fact, but never
realised it before.  The natives wear fur mittens, and I only wish that
I had possessed a pair yesterday.  Did you ever hear of ear-caps?
I never did until my last visit to Concord.  They are made of fur
or double velvet, are slipped over the ears, and held in their places
by an elastic band under the chin.  On windy days these are a
necessity for those who ride or drive.  The cloth travelling-cap is
nowhere, and I have been obliged to substitute a fur cap, as in
general use here, the wind going through the cloth as though it
were gauze.

Another institution is dust.  It is not black, because there is no
smoke, but is pale grey, and so impalpably fine that it penetrates
everything.  Yesterday the storm-sashes were put into my window.
In spite of the double window, and the great height of my room
from the ground, yesterday's wind drove the dust into the room so
fast that I was obliged to keep a clean handkerchief by me, and wipe
the paper continually.

This morning the wind has gone down, but the cold is still
intense.  When I came back from the early service I could not
open my mouth, the moustache and beard being frozen together.
And when they were thawed my lips were so stiff that I could not
speak to order breakfast.  On looking at the official record, I find
that the temperature was 5° *below* zero. . . .  I had to cross
the common, which is exposed to the N.W. wind.  Never was I so
thankful for fur.  The collar, when turned up, covers the ears and
both sides of the head, coming well above the eyes.  Even with this
protection my ears felt as if they had been frozen, and could be
blown off in fragments.  And, though I kept my gloved hands in
my pockets, they felt frozen.  This sort of weather is called a "cold
snap," and is mostly followed by a thaw, or, as they call it, a "warm

spell." This has already begun. At 4.30 p.m. I walked over the frozen ornamental water. At 6.0, on crossing the road, my boots were covered with *mud.* After the " warm spell " snow is expected. But this is an altogether abnormal year.

The expectations of the weather prophets were speedily realised, for on the very next day there is the following entry in the " log " :—

Here *is* the snow, and the boys, who have been fretting over their sleds, will be pleased. Sleds are stacked in bundles everywhere. They are painted of the most brilliant hues. They are shaped like swans, dogs, beavers, skunks, &c. There are double and tandem sleds. When there is a certain depth of snow the mayor issues an order prohibiting wheels, and all traffic has to be done on runners. The carriages are so made that they can be taken off the wheels and put on the runners.

December 17th.—I should just think that there *was* snow. It is deep enough to paralyse London, and is still falling. In spite of the double sashes it found its way through the window, and when I got up at 6.30 my table and books were covered with snow. So I turned to at once, put on some clothes, and dusted the snow off without difficulty. In London we do not prepare for snow, but here the people do prepare for it, and traffic is not stopped. Large gangs of men are at work, the sleighs are dashing about, and it seems so strange to see the traffic and hear nothing but the ceaseless jingle of sleigh-bells.

December 23rd.—The snow is unexampled, even in Boston, and all the appliances can scarcely keep pace with it. First came the snow-plough, looking like a railway-truck with a great ploughshare on each side, and diagonal cylindrical brushes beneath. It is drawn at full speed by six or eight horses, and is covered with men yelling and shouting just like firemen. It stops for nothing. Only yesterday two men were killed by one of these ploughs. Behind it come the car ploughs. These are much smaller, but exceedingly heavy, and are furnished with scoops on each side so as to clear the rails of the tramways. Then a legion of men are busy with ice-shovels, clearing the snow from the fronts of the houses, and throwing it on the banks raised by the ploughs. Then come a number of great snow-carts, into which a second legion of men pitch the snow which has been thrown up. This year the snow was too much even for Boston—one hundred and ten hours of continuous snow having taken place.

There was a very pretty turn-out on Wednesday evening. The sleigh was very ornamental, and was drawn by four splendid horses. Four strings of bells were hung round the body of each horse, a bell necklace on its neck, and bells of larger size hung like fringes upon the pole and harness. They were beautifully tuned to each other, the tinkling of the horse-bells harmonising with the chime of the harness-bells. Some sleighs are made like vast canoes or ships, and can take a whole school at a time. They carry banners and music, and the heads of the eight horses are decorated with brilliant plumes.

You should see young ladies going to school. I saw the whole process of the preliminary attiring. First the girl puts on a pair of jack-boots, reaching to the knees. These boots are indiarubber. Then she slings her books, papers, and luncheon-basket over her shoulders. Then she puts on a fur-cap, with the side flaps and peak let down. Next she puts on a waterproof sleeveless coat, pulls the hood over the cap, slips her hands through the coat, and fastens it inside. Lastly, she pulls her fur gloves through the openings, puts them on inside the coat, and then starts off. You meet numbers of them in this costume about 9 a.m. and 5 p.m. I put down all these details while they are new to me, knowing how soon they will be familiar and attract no notice.

Drivers of covered carts are protected by a vast waterproof screen, which is fastened to the front of the cart, and drawn nearly to the top. The reins pass through slits in the screen, each slit being covered with a flap, like those in the waterproof cloak. As the cart approaches you can see nothing of the driver but the top of his head and his eyes. The horses are protected by a waterproof covering, almost exactly like the housings used in the tournaments of the middle ages. You see nothing of the horse except the eyes and the feet. These screens and housings, however, are only used in carts and waggons, where the pace is slow and there are long halts.

Most thankful I am for the furs and "arctics." Without the

latter it would be almost impossible to walk, partly because the snow, where there is no householder to remove it, is several inches deep. Moreover, when the snow-carts cannot immediately follow the ploughs, and you have to cross the street, you will have to charge the snow-wall which is thrown up by the plough. Many persons wear boots like those already mentioned, and draw them over the trousers. Nothing but "rubbers" seem to hold the ice which gathers on the pavements. Even with them it is necessary to watch every step. As to horses, the poor beasts are tumbling down in all directions. As to humanity, even the arctics fail when ice forms on glass or iron slabs, and the boys have made slides on it. Yesterday, when returning from arranging the platform at the Cooper Institute, I found my feet slipping away from all control. Long experience of ice had taught me not to resist, and accordingly I sat down as gracefully as circumstances would permit. It was quite dark, and the "total depravity of inanimate objects" had put a glass slab *plus* ice *plus* slide in my way.

Another horse killed this afternoon. All yesterday the rain fell in torrents and produced a most curious effect. A dense white cloud rose from the snow, just as if hot water had been poured on it. Last night a short but fierce gale arose, followed by sharp frost. Consequently, the snow which had not been removed was first partly melted, and then froze, so as to make an uneven and slippery surface. . . . The way in which the sleighs charge the snow-banks is a caution. They go at them as fast as the horses can gallop. The sleigh is pivoted to the shafts in rather a curious manner, which allows of considerable up-and-down play, so that it shoots up one side of the bank and down the other without injury. It is rather startling to the novice, and makes him think that he is going up one side of a gable roof and down the other ; at least, it had that effect on me.

One effect of the dry cold is very remarkable. Here the barbers cut hair just as if their customers were convicts. Consequently, my hair looked much like a tight sealskin cap. It has now grown a little, and during the damp weather was perfectly manageable. But this cold dry wind has an electrical effect upon the hair, so that it sticks up like hedgehog's quills. Every now and then I pass a wet comb through it, but the effect is not lasting.

In spite of the extreme severity of the weather,
however, so different from that to which all his life he
had been accustomed, my father suffered but little, and
really did not take cold more often than he would probably have done during an ordinary winter in England.
Throughout his transatlantic visit, in fact, his health
was excellent, the ·constant travelling did not produce
its usual bad effect upon him, and almost all that he
complains of was the enforced irregularity of meals,
especially during the long railway journeys, with occasional remarks upon the weak points of American
cookery.

While staying in Boston he made the acquaintance
of Mr. Henry Irving, who was then touring in the States
with the entire Lyceum company.   And very much he
seems to have appreciated his friendship, which took the
pleasant form of complimentary tickets for the theatre,
and invitations to supper afterwards.   Here is one of
his notes upon the subject:—

Yesterday, just as I had sat down to luncheon, H. Irving sent
his carriage for me, saying that a box was at my service for *The
Merchant of Venice.*  Of course the luncheon was let alone.  I tried
to find the S——s, but they were not to be seen, and I had to go
off alone; so I had the box to myself, and felt horribly selfish.
Fancy the S——s' feelings when I told them at dinner! They
propose to wait in the dining-room all day, in case another offer
should be made.

There was one most impressive scene.  Shylock's house is on a
canal, and just beyond it is a bridge, under which gondolas pass
continually.  After Jessica has robbed her father under cover of the
masquerade, the revellers rush and dance across the stage.  Just
after the disappearance of the last masquerader, Shylock is seen,

bearing a lantern, on the opposite side of the bridge. He crosses it, goes to his door, and the curtain falls as he stands knocking at the empty house.

Portia was admirable, and her swagger off the stage as Balthazar was, excruciatingly droll. The care which was taken with the smallest parts was evident to everyone who was familiar with the stage. Even the general public behind the barrier was as carefully dressed and drilled as any of the actors. After Shylock leaves the hall the public rushes after him, and you can hear their hoots and hisses following him and melting in the distance.

In the evening I got a note from H. Irving, asking me to supper at 11.30. Miss Terry was good enough to come from her own hotel, and was accompanied by two well-known ladies, Mrs. L—— and Miss S——. The manager of the company was there, together with another gentleman, whose name I did not catch. It was very funny. In the course of supper, mention was made of a rattlesnake as thick as a man's thigh. Naturally I suggested that there must·be a mistake, whereupon Irving said that probably *his* thigh was intended as the measure. Then there arose a discussion, in the course of which I showed the distinction between venomous and constricting snakes. They were all greatly taken with the description, and Irving said that I had learned a good deal about snakes from my father ! This led to an explanation, and utter astonishment fell upon all, mixed at first with evident incredulity.

Such misapprehensions, by the way, were by no means infrequent, and for many years towards the end of his life my father was constantly taken for his own son. Perhaps the funniest blunder of this kind was that of a lady who imagined that the Rev. J. G. Wood was contemporary with Goldsmith !

# CHAPTER XIII.

## THE FIRST AMERICAN TOUR (*continued*).

Success of the "Lowell" Lectures—The Sailor and the Whale—Interview with
Dr. Oliver Wendell Holmes—A Dinner Party of Professors—Some more
Amèrican Nomenclature—Various American Peculiarities—Shop-window
Advertisement on a large Scale—"Dangerous Passages" and their
Meaning—Gold *versus* Bank-notes—American Chairs and their Defects—
A Curious New Year's Custom—Pedestrians and their Difficulties—Super-
stition and Religion—The Cult of the Horse Shoe—American Railways and
their Peculiarities—House-lifting—The Hotel "Office"—"Checking"
Luggage—"Serfdom." *versus* "Freedom"—American Clerical Costume—
The Black Waiter and his, ways—Negroes and "Coloured Men"—A
Black Waiter with a Cold—Stories of Negro Life—Negroes in Office—The
Chinese in America—Wing Lee and his Troubles—Opium Smoking—
American Manners—Children and their Behaviour—Thanksgiving Day and
Washington's Birthday—Incessant Elections and their Results—American
Cookery—"Pie"—Bear and "Corn-cakes."

MEANWHILE the stipulated twelve sketch-lectures were
being delivered on Tuesday and Friday evenings at the
Lowell Institute, and always with great and increasing
success. Just as in England, the drawings excited the
greatest enthusiasm. People listened with deep interest
to the lectures themselves, and were always quick to
pick up a point, or to follow the thread of an argument; but the loudest applause was always evoked by
the rapid sketches which were made in so many colours
upon the great black canvas. I quote my father's own
account of the lecture on the Whale, the first part of
which was given on the 13th of November :—

Room more crowded than ever, and the rush for front seats tremendous. Three ladies, for whom I had contrived to obtain tickets, said that they had been to these lectures for many years, and that they had never witnessed anything like the rush. One lady had her bonnet knocked off, and more than one thought that there was an alarm of fire. When I opened the lecture by drawing the whale, eleven feet long, in two strokes, there was first dead silence, and then such a thunder of applause that I had to wait. When the little sailor was drawn a number of lads gave a shriek of laughter, setting the example to the audience, who laughed and cheered in the heartiest manner. After the lecture was over the platform was stormed by local science.

Perhaps I may be allowed here to explain that the drawing in question was that of the spermaceti whale, and that the "little sailor" referred to was placed upon the whale's back in order to show the comparative sizes of the man and the cetacean which falls victim to his harpoon. The sketch always proved a very popular one, and was invariably received with much laughter and applause.

The second part of the lecture, given three days later, was equally successful :—

Audience still crammed in the evening. At the outset, Dr. C——, the secretary, told me that the audience had become so critical, from many years' experience of the best lecturers who could be obtained, that an incompetent lecturer, no matter how able he might be in other respects, would find the room gradually deserted. However, exactly the opposite result has been obtained, the auditorium being one dense mass of human heads. Yesterday's lecture was no exception, and it is really pleasant to have the platform stormed by the best men of the day, all lamenting that another half-hour at least might not be allowed. One of the professors was good enough to suggest that I should go on for two hours more, and was met by a general cheer.

On November 14th an event took place which had long been looked forward to with great pleasure—viz., an interview with Oliver Wendell Holmes, the famous American author. My father had for many years been a great admirer of his works, and we had all shared in his admiration; and so the meéting was regarded with special interest, not only by my father himself, but by us all. He writes about it as follows:—

> Be envious, all of you. I have had a long talk with O. W. Holmes. He was more than cordial, not to say affectionate, and spoke in the highest terms of my books, especially of the "Natural History of Man," which he characterised as an "encyclopædia of anthropology." He gave me another signed photo [one such had already been sent by him], and signed two more which I brought with me. Such a lovely view from his study window. It over-hangs the bay, which is called "Charles River," but which is really sea, and which comes within four or five yards of the window.

My father met the doctor again before very long, this time at a dinner party of professors given in his honour by one of the leading publishers of Boston.

> I was next O. W. Holmes (he writes, on December 1st), who was in high feather. Only fancy, he and I were the only two who had read "Little Pedlington" and "Uncle Remus." He is de-lighted with "Brer Rabbit," and especially so with "Tar Baby." It was a wonderful evening. No one was in the least dignified, and the chaff was merciless, and spared no one. There was a delightful German professor of natural history, whose description of a baby elephant was excruciatingly funny, especially when the baby in question "did walk from os, wearing ze preeches ob his fader."
> There is nothing like cheek. Next day Dr. Holmes called on Mr. H——, and said that he had enjoyed the evening immensely. Because, as he said, all the talking was done by the old men

—*i.e.,* Drs. G——, H——, W——, N——, and J. G. W——; while the young ones—*i.e.,* himself and Mr. H——, sat quietly, and improved their minds with wisdom. Now, as Dr. Holmes talked more than all of us put together, this was not a bad specimen of cheek.

Scattered throughout the "log" are curiosities of American nomenclature, some of which I give herewith. The ordinary hackneyed expressions are altogether omitted.

Did not you think that a drummer was some one who beat a drum? Here the word signifies a commercial traveller.

I always thought that a crank was a part of machinery, but here it signifies a lunatic.

A big advertisement (pronounced "advertízement"), or a candidature for office, is called a "boom." If it fails, it is a "fizzle"; if it succeeds, it is a "bonanza." A "bonanza girl" may mean an heiress, or the reigning beauty, or a famous horsewoman, &c., &c. "Museum" has the accent on the first syllable. What we call a railway rug is here called a "lap robe."

Another curious expression : to "enthuse." This is the electioneeringest country in the world. No sooner had they elected the governor than they began on the mayor, who is here elected by public vote. There was a procession last evening nearly, if not quite, two miles long, with brass bands, drums and fifes, and American flags by the dozen. I was asking what it was all about. My interlocutor was quite disgusted with the tameness of the whole affair, and said that "Borston didn't enthuse worth a cent."

To "beat" your way to a place signifies to evade payment of the fare. To beat hotels means to go into these establishments, eat a big dinner, and run your chance of being kicked out. A "tintype" signifies a hardly perceptible touch of negro blood in an apparently pure white. It is sure to show itself, either by the peculiar negro intonation ; by the absence of the "half-moon" on the finger-nails ; by involuntary swinging or jerking of the body or limbs ; by a sudden burst of laughter or irascibility without

apparent reason ; or, last of all, by the odour. It is quite perceptible even in the octoroon.

Doorsteps are called "stoops," and the pavements are "side-
walks." A "bobtail" is an omnibus with no conductor. Collared
head is "head-cheese." A "wild-cat" train is one that is not on the
passenger list, but is forwarded in the intervals of passenger traffic.

Certain other American "peculiarities" also come in
for notice :—

I wonder by what perversion of language the Americans called
side-spring boots by the name of "Congress Gaiters." A watch-glass
is called a "crystal." Ice is left at your door just like milk or the
newspaper. At first I was rather startled to see big blocks of ice
standing at doors, but am now used to it. The plan has the dis-
advantage of making the doorway slippery. The carrier dumps
down the block any way, and chips fly about, freeze to the pavement, and are horribly treacherous, especially in coming down the
outside flight of stone stairs which is universal here, the ground
having been reclaimed from the sea.

Shop-window advertisement is noteworthy. One huge estab-
lishment has got up the story of Red Riding Hood in wax figures of
life-size, with cottage garden and interiors to match. It begins with
the mother sending the child with her basket, and ends with the
grandmother's parlour (tall clock and all complete, with a view
through a partly opened door into the bedroom). In it is the
woodman, with his axe, the dead wolf on the ground, and Red
Riding Hood grasping the man's arm.

DANGEROUS PASSAGE.—This placard meets you everywhere.
On my first arrival I saw the notice in a small street. I looked
down the street, and as everything seemed all right, I went through
it to see where the danger came in. Presently I found another
Dangerous Passage which was quite safe, and then saw plenty of
them. Traffic, on foot or by wheel, passed through these dangers
just as if no warning were given. At last I found out what it
meant. The public streets are stringently regulated, and if anyone
be killed or injured the State of Massachusetts pays damages. But
the State declines responsibility for private roads ; so, the owner of

the property would have to pay damages. Therefore he protects himself by putting up this notice, so that anyone who passes along the street does so at his own risk.

I was under the delusion that Yankees loved GOLD. Whereas, they don't like it, and, when they see it, don't know what to do with it. I had a five-dollar gold piece which quite haunted me. People to whom I offered it looked at it suspiciously, turned it over and over, called their companions, and held anxious consultations over it. At last I really thought that I should be asked for my baptismal certificate. So I got it exchanged at the hotel for five of the ragged, dirty, evil-smelling "green-backs," which look as if they had lain about in the streets for a week or two of bad weather, and then been used for lamp-cleaning.

Chairs are a great trial to me. You can scarcely find a chair without arms. Then, they are so low that you can't sit up to the table with them, and they are so long in the seat that you *must* lean back in a semi-reclining attitude. Then there is the all-pervading rocking-chair. I am always tumbling over the projecting rockers. Even the revolving library chairs are fitted with a rocking movement immediately below the pivot.

A curious New Year's custom—probably designed as a sort of advertisement:—

On New Year's Day the hotel-keepers have a free luncheon for their customers, and vie with each other in "splendaciousness." Ours is said by the papers to have come off second. The first was at Hoffman House, Fifth Avenue, and nothing else has a chance against it. The proprietor struck out a most daring scheme, and made his luncheon-room into an art gallery, having nothing but the very best of everything. The room alone cost between five and six thousand pounds. It is surrounded by pictures of old and modern masters, among which is a Narcissus by Correggio, which simply blazes in its rich deep colouring. . . . So no one dreamed of competition. But ours *was* a luncheon! There was a *paté-de-foie-gras* so big that at 11.0 p.m. at least a quarter of it survived. I was awfully sorry for the geese, but the *paté* was something wonderful. It was put on the table at 1.0 p.m., and every one who frequented the house for billiards, luncheon, &c., might eat luncheon for twelve consecutive hours, if he liked.

A decided impediment to walking lies in the habit of putting great boxes of heavy goods on the side-walk. A huge waggon (drawn by five horses, three abreast in the shafts and two in the traces) pulls up opposite the store—say a sewing-machine business—and some 20 or 30 big square boxes are dumped down anyhow on the pavement, and the foot-passenger has to dodge his way amongst them if he can. If he can't he must go out into the road (already narrowed by the waggon) among horse-cars, "bobtails," buggies, carts, omnibuses, &c., &c., and so work round the obstacle. I find that the best plan is to get on the platform of a horse-car. Then the conductor explains objurgatorially that you are going in the wrong direction. You argue the point until you have passed the obstacle, and then you yield it and get off.

The following is curious and interesting :—

One would have thought, that the U.S.A. would be, of all civilised countries in the world, the least superstitious. Yet "Cranford," before the advent of the railway, does not yield to Boston or New York in point of superstition. The artisan, who will deride the Bible, class the clergyman with Mumbo-Jumbo, and deny a future state, believes with all his soul (only he says that he has none) in the HORSE-SHOE. Should he accidentally (he must not look for it, or the spell would be gone) find a cast horse-shoe, he is a happy man. He polishes it, he will go without his dinner for a week to gild it, he nails it over his door, and makes it his fetish He jeers at the very name of God. He yells blasphemies that are enough to curdle one's blood. But when you ask him about his horse-shoe he speaks with bated breath, and would, if he dared, shoot or stab anyone who would apply to it the same epithets that he applies to the Bible. This is not second-hand. I have seen it myself in factories where even the heads of departments could not feel secure without a horse-shoe over their doors. Let them have their fetish if they like. But it does seem the acme of illogical absurdity for a man to inveigh against the Bible as superstitious, and then to feel himself guarded against all earthly ills because he happens to have found a horse-shoe !

I cannot get over the calm manner in which express trains run through town without the least protection. The rails are sunk level with the ground. At one place the cars stopped in the middle of the market-place without even the vestige of a platform, and the passengers walked across several sets of rails in entering or leaving the train. I nearly lost my train at Worcester for this reason, Naturally, I expected it to draw up to the platform. O dear no! It *had* been standing for some time in the middle of several lines, and I had been looking at it without the least suspicion of its object. Finding casually that it was the Boston train I made a rush for it, and clambered on the car while it was in motion.

I saw a house being lifted, so as to get an additional storey below. Strong beams were inserted under the foundations, and a number of screw-jacks raised it enough to have wooden blocks inserted. Then it was again raised, and more blocks added.

The hotel office is a capital institution, and I wish we had it in England. It pays your cab when you arrive. When you go it procures your ticket and checks your luggage to anywhere, so that you need not trouble about it. If the journey be a long one, the office gets you a chair in a drawing-room car, and charges one dollar. It finds you a cab, and all the charges are entered in your bill. If you make purchases, you want no money. The articles are sent to the hotel, and the office pays for them. If you go out, and have forgotten your purse, the office hands you any sum you like.

The checking system is wonderfully good. When I neared Boston from New York, a man came into the car, bearing on his arm a huge iron ring, on which were hung numbers of leather straps. The man calls out "Any luggage to be checked?" I call out, "Hotel Brunswick; four." I give the man my New York checks, and in return he gives me checks marked "Brunswick." At the station a number of hack men are penned behind a barrier, and bawl for custom. These men may not come out unless called, but they may bawl and gesticulate as much as they like, and in the semi-darkness—glass roofs being impossible on account of the snow—all you can make out is a row of mouths opened beyond all anatomical proprieties, and double the number of hands gesticulating wildly in the obscurity. I call out "Brunswick," and the Brunswick driver takes possession of me. Arrived at the hotel I give the checks to the office, and shortly find

O

the luggage in my room. So you may travel from one end of the States to the other, carry nothing but a hand-bag, and never bother yourself about luggage. But there are no porters at the stations, so that any small baggage you must carry yourself.

It's an odd country in some things. In Maryland, no clergyman or minister can accept a legacy without permission of the Governor ! I prefer our "serfdom" to their "freedom." None of the clergy here, or in America generally, wear clerical attire, except when on duty. Even Dr. Phillips Brooks, of Trinity (who preached at Westminster, St. Paul's, Osborne, &c.), wears a black tie on week-days. I found myself as much an object of attention as if I had perambulated Regent Street in a surplice and biretta. So I was perforce obliged to conform to custom, and have modelled my costume on that of the Episcopalian clergy. And, though no one may believe me, I do *not* look a cad. The correct thing here is a black "sacque" coat (a sort of hybrid between a coat and a cutaway), waistcoat buttoning nearly to the throat, and filled in above with a broad black silk tie, so that the shirt is wholly covered; collars coming nearly round the neck, and nearly covered by the coat.

The black waiter comes in for a good deal of notice, and seems to be a somewhat comical individual. Here is the account of him given in the "log" :—

The black waiter is a queer being. He has not the least notion that time is of any value, and cannot walk straight across the room because he must diverge first to one side and then to another, and wait in the middle to swing his legs and flourish his arms. The other day, while walking down Tremont Street, I noticed in front of me a very fashionable young woman. Presently she swung her right leg round, and announced herself as a negress. The waiter can't keep himself still. If he hands you a visitor's card on a salver, he is obliged to keep the card down with his thumb, knowing that he can't help whirling the salver round as he presents it, just as a girl whirls a ball before throwing it. The other day, while bringing the salt-cellar for which someone had asked, the waiter chucked the salt high in air, and as it descended in a shower on the table was so frightened that he followed suit with the salt-cellar, which fell upside down on the heap of salt. If he carries a bunch of forks or spoons across the room he is sure to play bones with them unless the head-waiter's eye is severely upon him. However, I can get on very well with the negro, but the coloured man is a nuisance. If you look at a negro he grins, but if you look at a coloured man he sulks. In fact, all the coloured men are sulky in proportion to the amount of white blood in them. As to the genuine article, I never saw dignity till I saw a black preacher (of the "coloured Gospel") on Friday. He was one of the biggest men I ever met. He had the glossiest clothes I ever saw, and the shiniest hat, and the varnishedst boots, and the largest expanse of white shirt-front. Cetewayo, with a negro face, might have looked about half as majestic, especially if he were waiting for a horse-car. On Monday he will appear in striped jacket and apron, or in blue cotton overalls. As for me, I collapsed into nothingness. . . .

December 1st.—Black waiter again. One of them had a very bad cold. As he had often waited on me I compassionated him, and made him a glass of boiling punch in Oxford fashion. The head-waiter's eye was on him, but his excitement during the mixing process, in spite of his struggles to keep quiet, was nearly irrepressible. However, he took the glass and got safely out of the door, with only a low chuckle. Presently he came back, dodged into a corner out of range of the head-waiter, and there executed a dance, rubbing his gratified stomach the while.

December 23rd.—Mrs. P—— came out strong, not only telling, but acting stories of negro life. Once, on returning home from a rather long visit, she was met by her maid Lucy. "Heigh, Miss Phœbe! Glad to see you agin at de ole place. Is you bachelor woman still?"

Then there was a preacher of the "coloured Gospel" who besought "de Lord to look down wid sang-froid on de sins ob dem poor white trash."

When the negroes were suddenly emancipated they ejected all the scholars and gentlemen from the offices of State (this was in one of the Southern States), and put in a lot of absolutely ignorant plantation negroes. Lucy had married a fellow-servant named Pete, who was elected Governor of South Carolina. Of course they made such havoc that they were glad enough to go out and vote the white

men back again.  Sancho Panza was nothing to the negro in office.
Some time afterwards, when Mrs. P——, who had been married for
several years, took heart to re-visit her old home, Lucy found her
out, and welcomed her heartily.  Then she narrated her experience
of official life.  " Yes, Miss Phœbe, we lived in de State House, me
and Pete, and we gib big parties—heigh!  An, Pete, he sit at de
foot ob de table, and I sit at de *head* ob de table, Miss Phœbe.  An'
Pete, he toas' de ladies, and I toas' de genelums, Miss Phœbe.  An'
we had ducks for supper, Miss Phœbe, an' we drink champagne!  But
laws, Miss Phœbe, we was nuffin but niggers all de time."

The Chinaman, also, in his turn receives tolerably
frequent mention.  By reason of an old and somewhat
intimate acquaintance with Chang, the famous Chinese
giant, with whom at one time he used frequently to
exchange visits, my father took a special interest in the
nation, whose members consequently attracted more of
his attention than otherwise probably they would have
done.  The first note on the subject is quite a short
one, and shows the Chinaman hard at work in the
laundry.  Most of the washing in the larger American
towns, by the way, seems to be done by Chinese :—

November 24th.—Coming down Shawmut Avenue, I became
aware of an extraordinary gabble.  On coming near the house whence
it proceeded, I saw over the door—WING LEE, Laundry.  It was
a Chinese house.  I should think that some twenty Celestials were
in the room, which was a small one ; and the temperature was 70°
in the open air.

Poor Wing Lee afterwards came in for sad
trouble :—

You recollect my mentioning Wing Lee when I first came here.
The poor fellow was found in his laundry nearly dead from wounds

and bruises. It is believed that the assailants were his own country men, professional rivalry running very high. In the State of New York a law has been passed which forbids any more Chinese to land, and if a Chinaman goes out of the State he cannot return unless he has previously obtained a permit. In New York City no Chinaman may conduct a laundry within three blocks of another. Even Chang had great difficulty in getting leave to land.

There is a passage, again, with regard to the national habit of opium smoking, which, in spite of restrictions, appears as rife among the Chinese in America as it is elsewhere.

What wonderful rubbish the missionary world tolerates and talks. O! the virtuous indignation at the iniquity of England in *forcing* opium on guileless China! But, the same childlike and bland Chinee is smuggling his beloved opium into America, *he having grown it in China.* When a ship arrives at the quay, Ah Sin, a sailor on board, heaves a rope to Lee Fung ashore. Lee Fung makes the rope fast. After dusk he cuts the loop at the end of the rope, tucks it under his garments, makes a fresh loop, and goes home. The loop was nothing but an india-rubber tube full of opium.

Wing Lee does a great business in milk, butter, and new laid eggs. The eggs have all been emptied, and filled with opium. Foo Chang carries ice, a block being slung at each end of a stick which goes over the shoulder. The stick is hollow and contains opium. They *will* have it, and I do not believe that the Mongolian race is injured by it. On the contrary, when used temperately, as it is by nine hundred and ninety-nine in every thousand, it is beneficial. But we only hear of the one who abuses, and not of the nine hundred and ninety-nine who use it. I believe that the Caucasian and Hamitic races are not intended to find a solace in opium, which clears the brain of the Malay or Chinaman, but muddles that of the European. Of course excess would be injurious in any race.

N.B.—How can England, or France, or America, *force* a trade upon China, or any other country? Even Sir R. Alcock has come over to this opinion.

American manners, whether of adults or children, come in for scant praise. The boys are described as noisy and mischievous to the last degree, and absolutely wanting in all respect to their parents, who, strangely enough, seem perfectly blind to their manifold faults, and give them, if not open, at least tacit encouragement. Concerning the parents themselves my father has a good deal more to say; and he says it upon several occasions. As for example :—

I ought to have an angelic temper by the time of my return. There is scarcely anything which affects my nerves which is not in full play. The Americans are the noisiest people I ever met. They shriek at each other, and if *I* don't shriek no one understands me. The space at table which an ordinary American occupies is wonderful. Both his elbows are level- with his shoulders when 'he uses his knife and fork, and I get scrooged round the corner. When he is not eating he is whistling, or humming, or drumming on the table, or playing a tattoo with his knife and fork, or clattering with heel and toe on the floor. He kicks your chair at regular intervals, and you find yourself bracing up your nerves as the time for the next kick approaches. An American gentleman excused these ways by saying that the Americans were an excitable race, and not stolid like the English. I retorted that we were both of the same stock, and that the English were just as excitable as the Americans. Only, I said, that in England we are trained from childhood to refrain from such habits because they cause annoyance to others. I had him there.

If you are looking at a print in a window, someone will swagger along with his hands in his pockets and bump you aside. The first time this happened I instinctively started for the man's collar and the slack of his t . . s. But a moment of reflection showed that it was only boorish ignorance, and that he would have expected to be shoved aside in the same way himself. Or, the man thinks that he would like to look at the same print. So he leans on you, rests his chin on your shoulder, and clears his throat in your ear. He has not the least idea that it annoys you; it would not annoy *him*. Oddly enough, this pushing and bumping is far worse in Boston than in New York.

The great American anniversaries appear to have met with but very scant favour, to judge by the brief remarks passed upon them :—

November 29th.—This is Thanksgiving Day, and a horrid nuisance it is. Shops all shut, no late dinner at the hotel, and everyone supposed to be gorging themselves with turkey. Even the prisoners have turkey, and it is a point of honour among the well-to-do people to see that even the poorest family in their district shall not be without their turkey. That is a most admirable trait in the American character. Only in America there is none of the abject poverty of East London. Last year, at Concord, Dr. C—— had charge of a district (which means several parishes), and only found one family which needed a turkey. . .

Bother Washington! If he had never been born he would not have had a birthday, and that birthday would not have become a sort of bank holiday, to the upsettal of all arrangements. The cars were so crowded that I had to travel in the smoking-car, and even in that there were at least twenty who had to stand; of course stars and stripes and eagles were everywhere.

The constant elections, too, come in for severe condemnation :—

These incessant elections are the curse of the country, and I was told yesterday that "Presidential year" costs the country about as much as the whole expenses of the late war. Then, every year the Governor, Mayor, Marshal, &c., of each state are elected, and of course it is, as a rule, not the best, but the noisiest and most unscrupulous candidate who wins. The more thoughtful among Americans are already suggesting that the Presidential term ought to be at least seven years, and the Governorship of a State four years. Of course the Irish, being noisy and unscrupulous, are getting the government very much into their own hands. You don't find them in country towns or villages. They congregate in big cities, like Boston and New York ; and as three Irishmen make more noise than a hundred Yankees, the result may be imagined.

Then there is something about the American cookery :—

I have had supper at the Club, and had Tarrapin for supper. "Brudder Tarrapin" is very good, but is so rich that a little of him goes a long way. When you eat a bit of him your lips are glued firmly together. The meat is black and the bones are white, and there are too many of them.

Also I have had "pie." Never again ! The crust is much like the sole of an india-rubber shoe, and what the interior is like I cannot tell. A huge plate full of great slices of pie is put on the table. You look round and there it isn't ! Four days ago I took a very little piece, with barely a mouthful of crust, and am paying for it still. I don't wonder at the number of doctors. This morning I counted twenty-two doctors' plates on one side of one small block, most of which was taken up by the great Hotel Berkeley, and four or five lodging-houses.

I have been living considerably on bear lately. Saddle of bear is very like saddle of mutton, and there is rather more fat. It "sets" quickly on cooling, like venison fat, and so must be eaten very hot. Bear steaks are also very good, and so are bear chops.

"Corn-cakes" are very good, except that they are too dry to be eaten without butter, and all American butter is very salt—the cake being very sweet. Fresh butter is absolutely unknown. Is there not some way of squeezing butter, or putting it through the mincing machine, or rubbing it with the bread grater under water, or some such process, by which the salt is extracted ?

# CHAPTER XIV.

## THE FIRST AMERICAN TOUR (*continued*).

MEANWHILE the business of the tour was going steadily on. The last of the twelve Lowell lectures was given on December 7th, with the usual success, and previously to this four lectures (two on the " Entomarchetype," which had now been enlarged and extended, and two on " Ant Life,") had been delivered at St. Paul's School, Concord, North Hampshire. Concerning the opening sentences of the first part of the former lecture, the newspapers had something to say; and I find the following entry in the " log " :—

" A Boston lecturer is charged with using a dozen words in a single lecture that are not in any dictionary ; he alone knows what

they mean." This is *me*. I only began my lecture by explaining
that the "Entomarchetype" signified the typical arthropodous,
trachæiferous, hexapodous, annulate. Consequent consternation.
Then I told the audience that, if they wished, the whole of the
lecture should be given in that style, but that if they preferred
simple language they might have it.

On December 8th, "Pond and Stream" was given
as a private lecture before the members of the "Ladies'
Saturday Club" at Boston.  On the following Monday
came the "Whale," at Salem, Mass., which elicited the
following pretty little piece of criticism from one of the
leading newspapers :—

When I went to the lecture my feeling was : "What a pity he
has not a more interesting subject.  Whales?  I don't see how he
can say much that I shall care to hear about them."  But when
there, and when I came away, I wondered if anything else could be
so interesting.

My father's own account of the lecture, and of the
subsequent "Reception" which was held in his honour,
is as follows :—

December 11th.—Yesterday I lectured at Salem, and quite took
the place by storm.  The audience screwed an hour and fifty minutes
out of me, and then grumbled because the lecture was so short.
Afterwards, Mr. W——, my host, gave a "Reception."  Salem is the
scientific centre of America, just as Boston represents literature, and
Harvard education.  More than thirty of the leading men were
there, and they were all so taken with the lecture that they want to
get me to return and give a course.  Such a supper!  Game of
various kinds, fruits various, shell-oysters, oysters in patties, stewed,
fried, and done in all kinds of ways ; pyramids of ice ; champagne in
big glass jugs ; and everything in similar style.
    This morning, on leaving the building, I found a carriage for me,
with orders to be at my disposal all day.

On December 20th followed another lecture at
Boston; not at the Lowell Institute this time, but in
the Tremont Temple, for the American Society for the
Prevention of Cruelty to Animals. The subject chosen
was " The Hoof of the Horse," and the lecture seems to
have given rise to much subsequent discussion, and not
a little ill-feeling among the local farriers. The audience
themselves were remarkably enthusiastic in their expressions of approval :—

On the 20th (runs the " log,") my horse lecture came off with
enormous success. I had hardly spoken the last words when an
excited crowd surged on the platform from both sides, and converged
on me. They trod in the water basin, and upset it. They knocked
the water-bottle over, and deluged my table, spoiling the colours and
losing the small sponge. But they were madly enthusiastic, and I
could pardon them. Requests were made on all sides for my return
to Boston, and repetition of the lecture.

I heard one man delivering a lecture of his own. He got hold
of a hoof and a shoe, and argued that as the hoof was rounded and
the shoe was rounded, the hoof was made for the shoe. Of course,
this was " spoke sarcastic."

Those personally interested in the orthodox system
of shoeing were far otherwise affected by the lecture, as
may be gathered from a subsequent passage :—

Here's a to-do ! The farriers and horse-shoe contractors have
risen in their wrath against me. One of them, who ought to have
known better, " went for me " in a lecture, and said that " such
shoes as I exhibited were the work of ignorant, beery brutes, and
could only have been found in England." That very man was at
my lecture, and had not the courage to attack me, though I invited
attack, and remained on the platform nearly half an hour after the
lecture. But, although he talked in that way, and wrote similarly
vitriolic letters to the local papers, he did not contradict a single

word that I said. The fact is, that not only does he hold a contract to shoe the tram horses, but he is a politician, and holds an elective post. Next year he will have to vacate, and start a "boom" for re-election. And he fears lest his opponent may say that he is not worth a cent, but was "knocked out" by a Britisher.

On December 22nd a visit was paid to New York, where a lecture was to be given at the Cooper Institute. The drawing-frame had been "expressed," and sent on before; and on arriving at the hall my father found, to his utter astonishment, that the janitor had succeeded in putting it up, and that all he had to do was to re-stretch the canvas, which no one but himself ever succeeded in spreading quite to his requirements. The hall was again crowded to overflowing, and the lecture a great success. And then there came a "recess" until the second lecture at the same institute on the 29th, which was spent at one of the New York hotels.

Here a rather amusing incident occurred. My father was always very fond of tea, and extremely particular as to the manner in which it was made; and, whenever possible, if he had any doubts as to the capabilities of the maker, he would brew it for himself. The hotel system he seems to have despised altogether. This is what he says :—

Hotel tea-pots are rubbish. They are very small and inconvenient everywhere, but here they are simply absurd. They are scarcely larger than a cricket-ball, and hold about half of an ordinary breakfast-cup. The handle is so small that you can only hold it between the finger and thumb. This, added to my damaged right hand, almost prevented me from using it. So, on Wednesday, I went out and bought a pint Japanese tea-pot. . . At breakfast yester-day morning I produced it. The effect was general consternation, and if it had been the boiler of a locomotive it could not have created a greater sensation. The waiters gathered round it, and gazed on it helplessly. At last one of them took it away. In about ten minutes he came back, put it on the table, and vanished. When I took it up it was empty. The man's intellect had failed to solve the problem. At last it was again taken away and returned—this time filled with hot water ! Fully half an hour was consumed before I got the tea.

To add to their perplexity, I had also bought (for ten cents) a basket tea-strainer, as slop-basins are here unknown. The intricacy of this complex piece of mechanism was too much even for the head waiter. However, he is a Frenchman, and cannot be expected to understand tea. No one here does. I have only met one American as yet who can make tea, and *she* doesn't know how to pour it out.

Again there was a week's interval, and then another lecture at the Cooper Institute ; and then a further week's interval, and a fourth lecture, which brought the series to an end. On the 15th " Ant Life " was given at Worcester, and was followed by " The Whale " on the 17th, " The Hoof of the Horse " on the 22nd, and " Pond Life " on the 2nd of February. These lectures, like all the rest, were most successful, but had to be given in a room above a " dime museum," where a number of Kaffirs were performing ; and in consequence they were a good deal interrupted by shrieks and yells from below. My father of course embraced the opportunity of paying the savages a visit, and found them very interesting. This is what he has to say about them :—

One of them was the best assegai thrower I have ever seen. He sent six assegais in succession into a circle scarcely more than three inches in diameter. The manager of the museum announced— 'This gentleman will now illustrate the use of the assegai, the weapon with which the Prince Imperial was killed." It looked rather incongruous to see "this gentleman" reading the *Boston Globe.*

A few native Indians were also seen upon two separate occasions :—

On Monday I saw some noble red men of the forest. Fact. They were the genuine article, gorgeous in scarlet blanket, deer-skin robes, leather leggings, moccassins, and any amount of ermine, eagle-feathers, bear-claws, &c. There were about twelve men and four women, and their presence in a railway-car was incongruity itself; of which they were serenely unconscious. They stalked along the platform after someone who was acting as guide. Then, at a distance of some six or seven paces, came the women. Their object, I believe, was to pay their respects to the newly-elected Governor of the State at the State House. . . .

I saw more red men at a station. They were not gorgeous, and they wore boots, which scarcely harmonised with feather plumes on the head. I think that they were some of the Gay Head Indians, who behaved so well at the wreck of the *City of Columbus.*

On the 29th of January the first of a further course of six lectures was given at Boston; this time at the Chickering Hall. "Ant Life," in two parts, was followed by "Pond Life," also in two parts; then came the famous "Horse" lecture; and, last of all, "The Great Sea Serpent."

This last was quite a new lecture, and was the outcome of careful investigations made during the first part of this same American visit. But for many years my father had been a qualified believer in the sea serpent. He did not hold that a serpent of gigantic dimensions was roaming the seas, to be caught sight of now and then in widely distant parts of the globe.

He did not credit the various tales of its speed, its ferocity, and its horrific aspect which occasionally appeared in the periodical press. But he *did* believe that all these different stories, emanating from so many different sources, must have some basis in fact, and that they could not be entirely due to the imaginative power of a number of independent writers. And soon after arriving in America, he happened to meet with information which greatly confirmed him in this view. First of all he saw a skeleton, which was reported to be that of the monster in question, and speaks of it as follows :—

Yesterday I saw the sea serpent ! At least, I saw his skeleton, 65 feet long. It has been so knocked about that only the vertebræ remain, and they are in a dilapidated condition. One of them was better off than the rest, and just in sufficiently good preservation to prove that the beast was an eel-shaped cetacean, the spine being flexible up and down, and not sideways. This accounts for the appearance which has been noticed by everyone who lays claim to having seen the creature. Dr. W. F. L——, who sailed alongside one of these creatures, and made a sketch of it, lives near Boston, and I am going to appoint a meeting with him. . . .

November 23rd.—Dr. L—— called on the 20th and showed me the original sketch of the sea serpent. He is going to copy it for me. . . .

November 25th.—Yesterday I called on Messrs. Houghton and Mifflin, the chief Boston publishers. Saw Mr. Houghton, who asked me to write a paper for *The Atlantic Monthly*. I shall take the sea serpent as the subject.

Further information was soon forthcoming, and the promised article—entitled " The Trail of the Sea Serpent "—in due course appeared, my father stating his belief that the so-called fabulous monster is in

reality a long-necked cetacean of great rarity, nearly
akin to the whales, and probably perfectly harmless.
Then he thought of delivering a lecture upon the
same subject, and gave a short outline of his views
to some leading members of the Thursday Club,
Boston, in a kind of quasi-lecture. The result was a
numerously-signed address, asking him to deliver the
full text of the lecture at an early date, in some building to which the public as well as themselves might
have access ; and the lecture was accordingly delivered
as the last of the series at the Chickering Hall, on
March 13th.

After the second of these lectures, on January 31st,
came a most amusing experience, in the shape of that
curious American festivity known as a " Leap Year
Ball." I quote the description in the " log " :—

Last night I went to a "Leap Year Ball" at Salem. It was
held in the "Hamilton House," just like a provincial assembly
room. There were about 150 guests. It was great fun, the
privileges and duties of the sexes being reversed. The gentlemen
carried bouquets and smelling-bottles, and kept dropping bouquets,
handkerchiefs, and fans, so as to make the ladies pick them up.
Gentlemen could not cross the room unless accompanied by ladies,
while the latter roamed about at ease, and selected their partners.
There was a venerable judge, with the biggest bouquet of pink
rosebuds that the mind of man can conceive. A lively young lady
asked him for the next polka. The judge drew himself up, con-
templated the audacious individual with an air of impertinent
wonder, and replied that his mamma never allowed him to dance
with persons who had not been regularly introduced. After each
dance the lady led her partner to a chair, bowed, and went off.
Every now and then a gentleman would feel faint, and then his
partner had to put him into a chair, fetch iced water for him, and

fan him. The scene at supper was comical in the extreme. The gentlemen were seated on chairs round the room, while the ladies spread table-napkins over their knees, "so as to take care of their dresses," and brought them delicacies from the tables. It was a curious sight. All round the room a row of black-coated men, each with a table-napkin on his knees, while the supper table was surrounded with the brilliant dresses, flashing jewellery, and white arms of the ladies. Several of the former imitated the tactics of old dowagers, and contrived to be taken down to supper three or four times. The shrinking, timid modesty of the gentlemen was beautiful to see. At 11.30 a cotillon began, and continued until 2.30 a.m. It was a wonderfully pretty scene, the figures being a mixture of waltzes, minuets, lancers, polkas, &c., while every now and then the whole lot melted·into a gigantic ladies' chain, extending all round the room.

Part of it consisted in the distribution of "favours," of which there was ample store. After going a certain number of times round the room, the lady took her partner to the favour table, and then selected the most appropriate favour that could be found. Then, after another turn, the gentleman was allowed to select a favour, and to ask his partner's acceptance thereof. Consequently, colour began to predominate. One gentleman was gifted with a rattle, several artificial flowers, two or three huge rosettes with streaming ends, a metal peacock with spread tail, and a basket. A lady near me had a row of five big square pin-cushions suspended in front of her bodice. Somewhere about 1.30 the balance of things was restored. A number of gentlemen who were employed in business had slipped away, so the ladies constituted themselves Mormons for the nonce, and went shares, or thirds, in a partner.

On February 8th a trip was made to Southborough, Mass., where the first of five lectures was delivered in St. Mark's School. On the following evening came a private lecture on the Cockroach, to the members of the St. Botolph's Club at Boston. On the 12th followed the "Horse" lecture at the Chickering Hall, already referred to; two days later came the second lecture at

P

Southborough; and on the next night the lecturer was
at Salem, discoursing on "Ants of the Temperate
Regions." "Ants of the Tropics" followed on the
18th, and this was succeeded by "Pond Life" on the
21st, and that in turn by the "Horse's Hoof" on the
26th. On one of these occasions the lecture was preceded by a small luncheon party, concerning which I
find the entry "Kitten and grouse." Whether this
was some novel experiment in gastronomy or not, I
cannot say. Probably it was, as on a previous occasion
my father had feasted on "jugged " cat, and thoroughly
enjoyed it.

The 20th of February was occupied with "Spider
Life," at Southborough, and the 21st with "Pond and
Stream," at Salem. Next came "Ant Life," at North
Easton, on the 25th, followed by "The Horse," at
Salem, on the next day, "Ant Life" again at Newport
on the 27th (at half-an-hour's notice), "The Whale" at
Southborough on the 28th, and "Ants of the Temperate
Regions" at Andover on the 29th. All this time the
weather continued very bad indeed, and some of the
experiences of the earlier part of the winter were repeated
in an intensified form. Note after note upon the subject occurs in the "log," from which I quote the following short extracts :—

January 5th.—Brilliant sun, and clear blue sky. *But* the WIND,
*and* the COLD. I said that the thermometer went down. Rather!
I thought 15° below zero tolerably cold, but here is 30° below. Yet
the additional—or rather the subtractional—fifteen degrees seem
to make little, if any difference. In Minnesota 'it has been 43° below zero, and near Chicago twenty cars full of cattle were snowed in,

and the poor beasts have not yet been reached. Of course, they have all been frozen to death long ago. I told you of the wind of the 4th. At Fonda, N.Y., a woman went out to fetch some clothes off the line. She did not return, and when search was made, she was found in the branches of a pine-tree, into which she had been blown ; one leg and one shoulder were broken. Such weather has not been known for thirty years, and people naturally connect it with the scarlet and green sunsets. Yet I have suffered more from cold on the Crystal Palace parade than here, though the temperature is fully sixty degrees lower, and the wind twice as strong.

January 6th.—Still cold. There was a fire at Chicago, and several firemen were severely injured—not by the fire, but by the water, which fell on them and instantly froze, the temperature being 29° below zero. . . . . No one can understand this cold wave. Here we have the temperature of the Arctic regions, and are in the latitude of *Central Spain!* Even Florida has learned for the first time what frost means. Just now I am in an upper room of a hotel, warmed throughout by hot-water pipes. There is a large anthracite fire. The fur rug is over my knees and wrapped round my feet ; and yet my legs are shaking with cold. . . .

The streets look so strange. Ears are invisible, and those who do not possess ear-muffs, or " pantiles," or fur caps, are obliged to tie many-folded handkerchiefs over their ears. The sensation produced in the ears by a sixty-mile-per-hour wind, with the thermometer below zero, is remarkable. The ears feel just as if they were seized between red-hot tongs. Ditto the hands. My thick double gloves are of no more use than if they were made of gauze ; and even when held in the pockets, the frost grasps my hands as in a vice. The fur glove alone can withstand such cold. The first breath on going into the open air seems to fill the bronchial tubes and lungs with ice. Without the furs I should be a prisoner.

January 8th.—Here's weather again. This morning Mr. P——and I went to the club. Gangs of men were working with picks, ice-spades, ice-axes, &c., at the snow-ice wall which runs along the middle of Fifth Avenue. They were cutting breaches in it so as to allow people to cross. Before an hour had passed the weather suddenly changed, and a heavy snow-storm had begun to rage. It was still falling at 7 p.m. Now, at 10.30, it has changed to torrents of rain and a fierce wind.

March 9th.—Never was such weather known. This morning I went to the Advent Church through snow and ice. The return, being up-hill, took forty minutes, fifteen being a very easy average. I was engaged to take luncheon with Mr. R——; you can have no idea of what I underwent in keeping my appointment. After luncheon the weather put in another improvement—*i.e.*, torrents of rain, hail mixed with the rain, and a north-westerly gale, cutting your face with the hail as if fired at with a charge of No. 5 shot. All the R—— family begged me not to venture the walk home, but to stay for the night. Here was felt the want of the cab. However, I declined the invitation, and started off. How long the journey took I cannot say. . . . Several times I was brought to a standstill. It was impossible to go into the road, because it was one mass of tramcars, each with four horses, and now and then preceded by a snow-plough. The "side-walks" had their curved surfaces covered with a substratum of ice, on which was a superstratum of mixed hail and snow, the whole being the most hopelessly slippery surface that I ever knew. Even had it been level ground it would have been bad enough. But just imagine what it was. Sliding down the slope would fling you among the tram-cars, and not even the "arctics" could find a hold on a surface of semi-ice upon a subsurface of complete ice. No mortal could hold an umbrella, and even the regulation sealskin cap would have been blown away had it not been fastened under the chin. Add to this the fierce wind, hail-laden, spinning you round at its own sweet will, and slashing the hailstones into your face and eyes, just as if you were being whipped with small twigs. Several times I was really frightened, not being able to afford a broken limb, and sincerely regretted that I had not accepted the invitation to stay at Mr. R——'s. When at last I entered Quincy House I could realise Blondin's feelings after crossing Niagara. If any one had offered me £20 to make the journey again I should have laughed the offer to scorn.

Comment upon these extracts is needless. The only wonder is that the attendance at the lectures should always have been so uniformly satisfactory.

On Monday, March 3rd, North Easton was visited again, and the second part of the Ant lecture delivered. Next day came the same lecture at Andover, and two days later, "The Horse and his Owner" was given at Southborough. Andover was visited again on the 10th, when "Whales" formed the subject of discourse; and "The Sea Serpent," as already stated, was delivered in the Chickering Hall at Boston on the 13th. This brought the tour to an end, as far as lecturing was concerned.

My father did not leave America, however, until nearly a month later, being busily engaged upon articles for several of the magazines, which he wished to complete before setting out for England. He was, moreover, holding almost daily consultations with one of the leading publishers of Boston, with regard to the publication of an *International* Natural History upon a very large scale. All the arrangements for this were completed before the month was out, but the scheme subsequently fell through, owing to the failure of negotiations with the English publishers.

Several interesting passages occur in the "log" about this time. Here is one on "wooden houses":—

On getting fairly out of Boston, we find ourselves among wooden houses. Even the mansions of wealthy men are by preference built of wood. They look exactly as if they had come out of a toy shop and been magnified. The usual colour is white, with bright green blinds (outside) and red roofs. Now and then an æsthetic owner paints his house all yellow, like a sunflower. They look as if they would blaze up like shavings if a match were dropped within ten yards of them.

Of course, one would think that such houses could not be insured against fire, and I am sure that none of our English offices
would take them at any price. Yet, so contradictious are facts,
the offices are only too glad to get the insurances, and, when a big
fire occurs, the stone and brick buildings are burned to the ground,
and the wooden houses escape. They look dismally cold, whereas
they are rather *too* warm. I am writing this in a wooden house,
and though the country around is a sheet of snow, and a window is
open, I can hardly bear my coat. This is the dodge. First there is
a frame, across which are nailed row upon row of planks. This
makes the shell. Over the planks (which have been seasoned for
some three years after being sawn) is a layer of water-proofed felt, or
brown paper. Over that comes the covering, made of " clap-boards "
—*i.e.*, boards with overlapping edges. Inside the house is a lining
of plaster, and upon that is the paper. The chimneys are necessarily
of brick or stone, and there is a basement of the same material.

If you happen to want more rooms, nothing is simpler. You
screw up the whole house some ten or twelve feet from the basement, and put another set of rooms below. Household life goes on
just the same during the process. If you dislike your neighbours,
or your neighbours' poultry or ducks, you buy a piece of ground
elsewhere, put down a new basement, and then go, house and all, to
the new locality. If you are attached to your garden, you can take
it with you, trees, shrubs, &c., all complete. Now why do not we
do this ! Americans think nothing of it. A stone or brick house is
treated in the same free and easy style.

Then there is a note upon oranges :—

You don't know what oranges are. I did not. They are
picked ripe in Florida, packed very lightly, and sent northwards at
once. The packing which is necessary for a sea voyage would smash
them. In the first place, in size they are as a peach to a plum. In
the next, you must hold them nearly at arm's length when you peel
them. As you begin to strip the peel, the essential oil flies out in a
small white cloud. If it should strike your face, it has the effect of
a nettle ; and if it should get into the eye, it is as bad as cayenne
pepper. The white portion is very bitter, and you must get rid of

every particle. However, it comes off at a touch. Then you pull
the orange to pieces, cut off the edge of each piece, and pull out the
pips. *Then!!* It is all juice, very sweet, and always seems to
retain its coolness in the heated rooms of a hotel.

The "train-boys" come in for very unfavourable
mention :—

The train-boys are nuisances. They pervade the train. They
carry baskets full of books, or candy, or pea-nuts, or "pop-corn,"
and put a specimen by the side of each passenger. They wanted to
sell me for a quarter—*i.e.*, twenty-five cents—a book which sells in
the streets for ten cents. While going to Salem on Monday, I
thought that every traveller must have been taking absinthe. It
was only aniseed candy, which the train-boy had forced upon them.

There is also an amusing story describing some of
the dangers attending a theatrical enterprise in the oil
district :—

There is a play called "Struck Oil," which was very successful in New England, and the manager thought that he would try
it in the oil districts. It is a very sensational piece, with the
usual virtuous victim, the cold and calculating villain, the faithful
lover, &c., &c. The success was tremendous, and the treasury was
filled. Hotel accommodation was not to be had, and all the gentlemen of the company had to sleep in a loft. The manager was very
restless, and about 2.30 a.m. he saw the door pushed open, and a
couple of huge ruffians come stealthily in. They went round the
sleepers, comparing notes, and from their conversation it appeared
that they were dissatisfied with the punishment inflicted on the
villain in the last act, and had decided on thrashing him. Fortunately, they knew nothing about "making-up." One of them
thought that the manager was the villain, but the other identified
him as the lover—he, of course, not needing much of a make-up.
Finally they decided that "the cuss had scooted," and went away.

Then there is a short paragraph which reminds one
of Burns' famous lines :—

> " O wad some power the giftie gie us
> To see oursels as ithers see us."

I hear on all sides that both Matthew Arnold and I speak
with " a strong English accent " ! That is, I presume, we don't
shut up the backs of our throats and speak through our noses.
Boyd Dawkins is liable to the same defect, and so is Tyndall,
but in a less degree. Miss R—— told me that when she returned
from a long visit to England she was painfully struck with the
shrill, high-pitched voices of her country-women, and was horribly
afraid lest she should go back to the same pitch. They all seem
to scream, and make wonderful use of their lips.

There is also a note to the effect that at the leading
churches the choir-boys are all of English birth, no
amount of training being sufficient to eradicate the
shrill, nasal intonation of the American lads.

Finally there are one or two more Americanisms :—

Goloshes are called by various names. The ordinary golosh is
here called a " kick-off," to distinguish it from the " Arctic," which
is buckled. The " Arctic," by the way, is a waterproof *boot*, with
India-rubber sole, sides, and toes, which goes over the boots, and, if
wished, over the trousers as well. It is impossible to keep one's
footing in snow-time with a leather boot. I thought that it would
be horribly expensive, but it is cheap, the very best being only
eight shillings per pair.

" Rubber " is a generic term. South of New England, goloshes
are called "gums." One southern gentleman made quite a sensa
tion here lately. He and his wife were calling at a friend's house,
and the wife did not make her appearance with her husband. This
he explained by saying that she was " only rubbing her gums on the
door-mat."

As to " kick-offs," when you tumble on the side-walk, the height
to which the article in question can be projected is really amazing.

If it would only rise perpendicularly, one would not mind it so much. But, in accordance with the total depravity of inanimate objects, it mostly falls among the horse-cars, or on the opposite side of the road, where recovery is hopeless. On February 29th Tremont Street was literally strewn with goloshes, the owners of which had wisely left them to their fate.

Before leaving America my father was entertained by several of the leading societies, and various " Receptions " were given in his honour  Concerning two of the most important of these latter he briefly remarks that modesty forbids him to enter into particulars ; with regard to a third he is a little less reticent.   A newspaper cutting is pasted upon the " log," to the following effect :—

The monthly meeting of the Papyrus Club was held at the Revere House, Saturday evening. Professor J. G. Wood, the English Naturalist, and Mr. Paton of the New York Tile Club, were the guests, and each responded to toasts in their honour. After dinner exercises were contributed by Mr. Barrett Wendell, Mr. F. J. Stimpson, Mr. John A. Lowell, Mr. T. Adamowski, and Mr. Otto Bendix. There were about fifty members and guests present.

Then comes the following annotation :—

This is not quite correct. The " reception " was given to *me.* Mr. Paton belongs to an affiliated club in New York, the members of which interchange hospitality. Note the curious use of the word " exercise," which has evidently come down from Puritan times. Prayers, hymns, songs, speeches, musical performances, &c., are all termed " exercises." Any way, the president " exercised " his laudatory powers to the fullest degree, and, according to newspaper phraseology, I was " handsomely ovated."

On the 9th of April my father left New York on board the Cunard steamship *Servia,* amid many

regrets at his departure, and hopes that he would
shortly return.  He had, indeed, almost pledged himself to revisit America during the following winter, and
was even thinking, and thinking seriously, of taking
up his abode in that country altogether.  But this
last idea, fortunately, as I think, for himself, was not
destined to be carried out, and although a second transatlantic visit was paid according to promise, it again
took the form of a mere lecturing tour of only a few
months' duration.

The return passage was accomplished under far more
favourable circumstances than that in the *Cephalonia*,
although the sea was again very rough, and bad weather
almost incessant.   On Thursday, the 17th, Queenstown
was reached, and early on the following morning the
*Servia* was in the docks at Liverpool.

Had the tour been successful ?

It had, and it had not.   Judged merely by the
enthusiasm which the lectures had excited, its success
could not be doubted.   In every town which my father
had visited he had been greeted by overflowing
audiences, the papers had been full of his praises, and
he had received invitation after invitation to return to
the country for a more prolonged tour.   But, from a
financial point of view, it was scarcely, if at all, more
profitable than would have been a winter in England.
Throughout his life my father suffered very severely in
pocket through his utter lack of business qualities.
Anyone who would could cheat him.   He never set a
sufficiently high price on his work, and very often did

not obtain all, even, that he had bargained for. And in this American tour he made the grand initial mistake of not putting himself into the hands of one of the great lecturing agencies, upon the usual understanding that a certain fixed sum was to be paid to him under any circumstances, with a certain specified addition thereto if the receipts should exceed a certain amount. In that case he would have cleared his travelling expenses, and returned home with a fairly substantial balance. But, instead of doing this, he crossed the Atlantic solely on the strength of the Lowell lectures, which, though profitable enough as one engagement out of many, were certainly not sufficiently remunerative to ensure the success of the venture. It is true that after a time he *did* secure the services of a professional agent. But by some ill-fortune he contrived to select one whose whole attention was practically monopolised by travelling dramatic companies, while no stipulation whatever was made as to the payment of a fixed sum in any contingency; so that the lectures fell into the background, and were not "pushed" as they would have been in better hands. The consequence was that, on more than one occasion, no lecture was given for a week or ten days at a time. The cost of living in America is very much higher than in England, travelling expenses had to be paid, and, when all was taken into account, the financial results of the tour barely exceeded those which would have been forthcoming from an average winter season in England.

Of course the matter had also to be considered from

another point of view. The tour had yielded much
personal enjoyment (for my father had scarcely been
out of England before), and had added considerably to
his stock of bodily health; and he had also accumulated
a good deal of matter which afterwards afforded the
basis of many magazine articles. But, regarded as a
business undertaking, which in the main it was, the
tour had proved but a very qualified success, and the
expectations with which he started had by no means
been fulfilled.

Of course by the time that he arrived at Liverpool
the English season was over; and save for three
lectures at Yarlet Hall, near Stafford, towards the end
of May, and five at a small boys' school in Upper
Norwood, given at intervals during the summer term,
lecturing ceased until the beginning of the season of
1884-85.

# CHAPTER XV.

IT had been well for my father, in more ways than one, if he had rested satisfied with his first visit to America, without deciding to repeat it in the following winter. He would then have escaped much disappointment and much failure, to say nothing of the consequent absence from the English platform during two successive seasons. But several causes conspired together to bring about his return.

In the first place, he had in some degree pledged himself so to return, before leaving for England. Forgetting the proverbial love of the Americans for novelty, and forgetting that he himself at a second visit would be a novelty no more, he imagined that the

success of his lecturing in one season would necessarily
ensure, not only an equal, but a far greater success in
the next.   No doubt he was badly advised by many of
the Americans themselves, who, as personal friends,
wished for a speedy repetition of the visit on personal
grounds.   No doubt, too, the interested representations
of his incompetent American agent had great weight
with him, and he was led to believe that the financial
success of a second visit would not only be absolutely
assured, but would greatly exceed that of the first.   But
the greatest mistake of all was that of overlooking the
fact that the forthcoming season would be taken up with
the Presidential and other elections, that the Americans,
absorbed in their deep and burning ardour for politics,
would have neither time nor thought for anything but
canvassing and party meetings, and that, consequently,
even under the most favourable conditions, no measure
of pecuniary or other success could be expected.   And
so evil counsels prevailed, and a second transatlantic
journey was undertaken.

   This, however, did not commence until comparatively late in the year, and the earlier part of the autumn
was occupied as usual with lectures in England.   These
alone saved the season from total and absolute failure.
The first, on " Pond Life," was delivered at Coventry on
September 23rd.   This was followed by " Spider Life,"
at the Royal Naval School, New Cross, on October 3rd.
On the 7th of the same month the same lecture was
given at Tonbridge Grammar School.   " Pond Life "
came again on the 9th at Saffron Walden, where the

hall-keeper seems to have been troublesome. So, at least, I gather from the diary, where I find an entry to the effect, " Very crusty janitor, who dodges off work *because of sprain in back.* Query : in temper ?

Next day came the old favourite, " Unappreciated Insects," at Haverhill. " Pond Life," which about this time seems to have been a specially popular lecture, followed at South Norwood on the 13th; and on the next day the second part of " Spider Life " was given at Tonbridge Grammar School. Then came " Pond Life " again, twice, the first time at Stamford, on the 16th, and the second at Sheffield, on the following day. On the 21st, the first part of " Ant Life " was given at Dulwich College ; then came " The Horse," five days later, at Marlborough College ; and this was followed again on the 28th by the second part of " Ant Life," at Dulwich.

The Marlborough visit gives some idea of the hardships that occasionally attend upon a lecturer. Leaving home at 11.0 a.m., my father found upon arriving at Paddington that the train by which he had intended to travel had been taken off; so that he had to wait at the station for rather more than an hour and a half. Arriving at the College at five o'clock, instead of at half-past three, he made the far from pleasant discovery that the preliminary letter which he invariably sent a few days before every lecture had never been delivered, and that consequently no arrangements had been made for the preparation of the lecture-hall. This difficulty, however, was overcome, but with the inevitable consequence

that the lecture was delayed; and so, after delivering
his lecture, and taking down and packing the drawing-
frame, my father was unable to reach his hotel until
very nearly half-past eleven. Even yet, however, his
day's work was not completed, for he found a newspaper
reporter waiting for him when he arrived, and had to
give an abstract of the lecture for publication before he
could obtain any refreshment. Consequently,, it was
fully one o'clock before he could retire to rest; and at
a quarter before five he had to rise, in order to drive
over to Savernake in time to catch the early train to
London.

On the 1st of November, the now famous "Horse"
lecture was delivered at Brixton, in the Geological
Museum belonging to Dr. Chaning Pearce, in which the
original sketch-lectures of all had been given rather
more than five years previously. "Dolly," an unshod
horse belonging to the doctor, was exhibited in the
greenhouse adjoining as a living illustration of the main
point of the lecture; and as she had been in regular
carriage work for several months, over ordinary London
roads, while her hoofs were nevertheless in perfectly
sound condition, proof positive was given of the
lecturer's contention that shoes are not necessary, and
that horses are better, and less liable to accident and
disease, without them.

Then followed in quick succession lectures at Rossall,
West Bromwich, and Leeds; and then came a few days
spent at home, in the necessary preparations for departure. Then, on the 11th of November, my father

left Norwood for Liverpool; and on the following day he embarked for Boston in the Cunard steamship *Catalonia.*

No "log" was kept this time, as the novelty of the journey and subsequent experiences had worn off; and the only records of the tour lie in the diary—almost always of the most sketchy and scanty description—and in the few letters which have been preserved. From the former of these I gather that the weather this time was of a much more satisfactory character than upon the occasion of the first journey; so much so, in fact, that a large board was roughly painted, in order to act as a substitute for the canvas drawing-screen, and a lecture, on the Stickleback and the Great Water Beetle, given for the benefit of the passengers on the evening of the 17th. This was so successful, and so well appreciated, that a second lecture was arranged for the 19th; but, as the waves were then sweeping over the deck, and the lecturer would have been quite unable to keep his footing, it was postponed until the following day. As soon as the corner of the Newfoundland "Bank" was turned the wind lulled, and the lecture—Part I. of "Ant Life"—was duly given; and this was followed by the second part on the evening of the 22nd. And early on the morning of the 23rd the vessel anchored in Boston Harbour, after an unusually slow passage of eleven days.

On the following day a visit was paid to the agent, who promised great things, and undertook at once to send out copies of the syllabus, &c., to all the leading

Q

institutes—a task which ought, of course, to have been performed weeks before. This, apparently, he did without further delay, but took no particular pains to secure engagements. And he also, without enquiry, allowed himself to be entrapped into an arrangement which subsequently fell through, and which, in itself, was quite sufficient to ruin the success of the tour.

Yet the season began fairly well, and the prospects were apparently bright and promising enough. The first lecture was given at Newton, Mass., on the 1st of December, and was followed by the second part of "Ant Life" at Southborough two days later. The first part of this lecture, curiously enough, had been given during the former tour, on February 8th of the same year; and although four subsequent lectures had intervened at the very same Institute, the subject had not until now been concluded.

Next day came the first part of the same lecture at Peabody, Mass., in the Peabody Institute. This was delivered under circumstances not the most agreeable, as a fancy fair upon a somewhat large scale was being held in the room immediately beneath, and the audience was largely recruited from the "rough" element, the members of which indulged in much horse-play, and not a little noisy comment upon the lecture. On the following evening came "Pond Life," at Haverhill, followed by "Bee Life," at Southborough on the 9th. There is a characteristic entry in the diary concerning this latter lecture, and one which well illustrates my

father's peculiar knack of obtaining without particular
difficulty that which very few men would have succeeded
in obtaining at all.

Went to Southborough, and mistook train; telegraphed; had
oysters; then took next train, *and bribed officials to stop train at
Southborough!*"

Two days later he went to Peabody, for the second
part of the Ant lecture, and found the fancy fair
still going on in the room below the hall; and the
audience on this occasion is tersely described as "awful."
On Saturday, the 13th, the first part of the ever-
popular "Ant Life" was delivered to members of the
St. Botolph Club at Boston, and on the following Tuesday "Whales" was given at Cambridge, Mass., with
the usual success. Next day, by a curious coincidence,
came "Ant Life," Part I., at *Oxford*, where the janitor,
through officious over-zeal, contrived to break the
drawing-frame rather badly. On the 18th "Pond
Life" was given at Peabody, where the fancy fair was
over at last; and this was followed, two days later, by
"Whales" at Nantucket, where the janitor is described
as "an old noodle, who *would* not do what was wanted."
The weather had now become very severe, with the
thermometer at 10° below zero; but on the 26th, when
a visit was paid to St. Johnsburgh, it had further
sunk to—27°, without, however, diminishing either the
numbers or the enthusiasm of the audience. Tuesday,
December 30th, saw the lecturer again at Oxford,
whither he had repaired for the purpose of concluding

Q 2

" Ant Life," and next day " Pond Life " was given at Concord, Mass. And so came to an end the year 1884.

Up to this time the tour had been fairly successful. Fourteen lectures had been given in the course of the month, and although these were less remunerative than those which had been delivered at the Lowell Institute in the preceding season, they had yet brought in sufficient to leave a moderately substantial balance after the payment of all expenses. But now engagements became few and far between. Those that were secured were so badly arranged that very possibly a long railway journey had to be undertaken in order that a single lecture might be delivered; and it soon became apparent that upon the tour as a whole there would be no profit at all.

The principal reason for this failure my father gave in a letter from which I quote the following passages:—

At last I can tell you something about the Western lectures. They have been a complete swindle. A syndicate of three persons got up a scheme for a vast lecture course through the West. About ten lecturers were to be engaged, and sent on a circuitous tour, so as to avoid backwards and forwards travel. Payment was to be according to reputation. They asked H—— to supply five lecturers, each to give from thirty to fifty lectures in January and February. I was considered the chief of them, and payment was to be either £20 for each lecture, and pay my own expenses, or £15 nett. The whole idea was an excellent one, but unfortunately the three quarrelled, and the whole thing was thrown over. H—— was quite knocked over by the blow, and could hardly speak when he told me of it. Besides all his work, postage, &c., he loses from £160 to £200 in his commissions, besides the injury to the prestige of his house. I cannot tell you what the anxiety of the last three weeks has been.

Here was my sheet anchor suddenly torn away, just when it was most needed. However, H—— acted with a promptitude for which I had not given him credit, and at once sent out a shower of letters on his own account. One place—La Crosse, near Chicago—he has already secured, and he has received a number of applications from Chicago. To save time, he is conducting his correspondence by telegraph.

All these applications, however—if ever they existed —came to nothing, and the first eight days of the new year my father was compelled to spend in inactivity at Boston. On the 9th of January he started for La Crosse, hoping against hope that other lectures might still be arranged either in Chicago or some of the neighbouring towns. But in this he was again disappointed.

From St. Paul's, Minn., on his way to La Crosse, he wrote as follows :—

I'll just write what I can. I arrived here at 8.0 a.m., after travelling all night. Thermometer 4° below zero, and no one to meet me. Fortunately I had the name of the secretary in my pocket-book, but not his address. So I put my belongings into the baggage-room, went into the streets, and by enquiring at the best shops, found the address. He is one of a well-known firm of lawyers. There I found one of the firm. As usual, each of them thought that the other was to meet me, and so I was literally left out in the cold. However, he was very kind, and went with me to the hotel, where he engaged a room on the quiet side of the house. The other side stands on the platform of the railway station.

As far as I can see, I shall have to pack my frame, have supper, and get to bed by 11.0. Then up at 2.30 to catch the 3.0 a.m. train to Chicago, where I arrive at 3.0 p.m. Then a thirty-six hours' journey from Chicago to Boston. So I need to be somewhat tough to get through all the fatigue. Just think what thirty-six hours' incessant jolting and trembling of the car must be. Then the monotony of American travelling is appalling after you are

accustomed to the differences of customs.  In England and on the
Continent you pass through a series of pictures, every little village
having its individuality.  But, here, every place is like every other
place.  There are the same white wooden houses and green blinds,
the same wooden churches with wooden spires that look as if they
could be unscrewed, and the same lack of durability, as if they could
be swept away with a birch broom.  *Per contra*, the scenery, when
ice on the car window does not shut out the view, is often lovely,
and, owing to the difference of trees, quite unlike the scenery of
Europe.  Anyway, I am glad to have seen these big lakes and the
Mississippi; though, as to the latter, I only saw the frozen surface,
which was traversed by big horse vehicles.

I have put up the frame in the "Opera House."  The enclosed
letter from H—— shows that it would be absurd to linger in the
West on the mere chance of someone procuring lectures somewhere.
As it was, my literary reputation got me this lecture, and H—— had
nothing to do with it, beyond sending me the date.  I at once sent
an urgent telegram to him, and by the time that I reach Boston he
*may* have something.  Anyway I am fighting my best, and it is a
mighty tough fight.  For nearly a week I shall only have three clear
hours in bed, rough-and-ready meals snatched anyhow (*fancy dining
at* 12.0!), and all for nothing.  I cannot find H——'s letter, but it
was to the effect that there may be lectures in and about Chicago;
whereas I, who have been to headquarters there, know that there can
be none.

Tea-time; ticket-taking, &c. &c.

When the time for the La Crosse lecture arrived,
the thermometer had fallen to no less than forty degrees
below zero; but the audience was nevertheless full and
very appreciative, and the lecture a great success; and
next day the return journey to Boston—a very
fatiguing one—was commenced.  My father writes from
Boston on the 16th as follows:—

Here I am, back again, after *such* a journey (fifty-two hours!)
On Tuesday I could not get to bed until after midnight, and had to

get up at 5.30. It was fortunate that I did so, as otherwise I should have had no breakfast. Being anxious about my luggage, I went to the railway-station (thermometer still 40° below zero), and after much difficulty found the baggage-master. He could not understand either my ticket or the map, and expended fully five-and-twenty minutes before I got the checks. Then back to the hotel, packed the travelling-bags, took them down myself, and put them by the door so as to be ready. The train was to start at seven, and breakfast was announced for half-past six. But the breakfast-room was not opened until a quarter to seven, so that there were barely seven minutes for breakfast. Then I had to carry the two bags, &c., and find the train. They do not trouble themselves about platforms here, but the trains stand about anywhere in the road or market-place. There are no porters or railway-servants about, and the only plan is to board every train until you find the right one. This is not pleasant, when you have to climb the steep steps and are impeded by luggage. As far as Chicago, all was plain sailing. There I had to take a carriage across the city, and get upon the Lake Shore line. Now began a series of difficulties. I had been puzzled to know why the fare to Boston was so expensive. To La Crosse *from* Boston, including drawing-room and sleeping-cars, the fare was twenty-six dollars, *i.e.*, about five guineas. But the fare from La Crosse *to* Boston was thirty-two dollars, or nearly £6 10s., *without* drawing-room and sleeping-cars. The latter are 10s. 6d. each, and the former 4s. Both are necessary, as the ordinary cars have no accommodation for washing, and the wood fuel covers you with a fine black dust. Neither is there any drinking-water, and the hot and close atmosphere creates incessant thirst.

Then began the bother of the tickets, my route being a singularly complicated one, involving all kinds of changes. Just before arriving at Buffalo, where the sleeper tickets had to be procured, the conductor found that I should have to take a carriage and drive to another station. When we arrived, there were only twenty-five minutes in which to cross Buffalo, take ticket, and catch train. No hope of supper. There were just seven minutes to spare, so I ran across the road to an apparent restaurant. It was a "saloon," where eatables are not sold. However, the mistress very kindly gave me a roll, broke it, and put a slice of ham inside, with which I bolted back again. As it happened, there were oysters and milk in the car, but the roll came in handy afterwards. I could only get an upper berth. Then I found that about six o'clock next morning I should have to make another change. So I turned out at 5.15 (getting out of an upper berth is anything but a pleasant process), and got myself dressed and bags re-packed. This is a difficult business in a sleeper, as boots, rubbers, overcoats, umbrellas, hats, &c., are tucked away by the conductor in all kinds of holes and corners, the berths and curtains occupying the whole of the car except a narrow alley, through which you must sidle. Light, of course, is very dim. Exactly at six the train stopped, so it was lucky that I was up so early. Then I had to wander again among miscellaneous trains, and found the right one. At Albany I had again to turn out, carry my bags about a quarter of a mile, and then go up three flights of stone stairs. This was at 6·45. Here I learned that the train was due at Boston at 3.0, and that there was no chance of getting food. Happily, the roll and ham (a very little piece) came to the rescue, otherwise I should have had sixteen hours' fast. As it was, I just arrived here as they were closing the dining-room doors, and induced them to let me in privately.

The conductor of the last train explained the mystery of the route. The clerks get a percentage on the tickets, and so the clerk at La Crosse chose the most expensive and complicated route. Altogether the expenses of the double journey were seventy dollars. H——, however, would only take half his fee. He is still in communication with several places; but I cannot afford to stay here on the chance of getting lectures, and, unless I know of something within a week, I shall telegraph to you and come home by the next boat. I trust that it may not come to that, but it is necessary to look ahead.

H——'s programme was a very sound one. It was to procure half-a-dozen lectures at least in December, then reserve January and February for the thirty to fifty in the west, and then, while I was away, to arrange for the whole of March. He procured twelve for December, and was in correspondence for March. Then came the unexpected blow from Chicago; and there has hardly been time to complete arrangements, especially as proposals for January and February had to be declined.

The fee paid for the La Crosse lecture, I may here remark, was one hundred and twenty-five dollars, or

£25. From this had to be deducted seventy dollars, or £14, for the travelling expenses, besides the agent's commission, and certain items of incidental outlay ; so that the week of incessant travelling and hard work resulted in a profit of perhaps some seven pounds.

With the exception of the second part of " Ant Life " at the St. Botolph Club on the 28th, and the first part of the same lecture at Dartmouth on the 30th, this was the only lecture delivered during the entire month of January!

Still the agent continued to make his delusive promises, and still nothing came of them. Even if no other matter of interest were engaging men's minds, the time had long passed by for making lecture arrangements. But, moreover, political questions were everywhere wholly absorbing the public attention ; everywhere lectures, concerts, and even dramatic performances were being set aside in favour of the meetings of the rival parties, while no one could talk or think of any subject which was not in some way connected with electioneering. And so, when the first day of February dawned, and still no lectures were pending, my father accepted the inevitable, secured a berth on board the Cunard steamship *Cephalonia*, which had brought him over to America on the occasion of his first visit, and which was now advertised to sail for England on the 7th of the month, telegraphed to us to expect him at home on the day after the vessel reached Liverpool, and made all his preparations for embarking in less than a week's time.

Only one lecture now remained to be given—a private one, at the house of a friend in Charles Street, Boston. This was duly delivered on the afternoon of Thursday, the 5th, and met with great appreciation ; and two days later my father transferred his belongings to the *Cephalonia*, and left the American shores for the second and last time.

The tour, as a whole, had been a complete failure, the arrangements having been hopelessly mismanaged from beginning to end. It ought not, in the first place, to have been undertaken at all until several years at least after the first visit. It ought not, under any circumstances, to have taken place at a time when electioneering excitement, always so fierce in America, was in its fullest force, and when, in consequence, men's minds were certain to be in a state of more or less turbulent disquietude. Neither should the business arrangements have been postponed until almost the time of my father's arrival in America; nor yet should they have been entrusted to an agent so wanting in energy as the man upon whom his choice unfortunately happened to fall. Mistake, in fact, followed mistake, and as a necessary result, when the tour was brought to its sudden and premature conclusion, expenses had barely been covered, valuable time had been uselessly wasted, and profit there was little or none. Nor was my father alone in his disappointing experiences, for two others of our best known English lecturers were in precisely the same predicament. And all three returned to England almost together, and long before the time originally fixed upon for their departure.

The passage home was a moderately quiet and easy one, broken only by one incident of any note: that of a sudden outburst of madness, soon after the vessel had weighed anchor, upon the part of one of the female steerage passengers, who had to be forcibly restrained for some hours, and finally died in a paroxysm of mania upon the day after sailing. On the 17th the vessel touched at Queenstown, reaching Liverpool on the following day; and so the second of the two American journeys was brought to a close.

Of course, as the English season was very nearly at an end, but little could be managed in the way of engagements after my father's return; but, by the energy of his English agent, a few lectures were nevertheless arranged for within a few days after his arrival. The first of these, on "Ant Life," Part I., was delivered on February 25th at a large ladies' school at New Southgate, Middlesex, where many others were subsequently given. At this school, as at most of the many others at which my father lectured, guests from the neighbourhood were invited to attend, and the audience was consequently by no means confined to the pupils alone, but was of very much the same character as that which would have listened to him at a public institution.

Next day "The Whale" was given at Clifton College to a very appreciative audience of boys and masters; and on the following evening the same lecture was delivered at the Seamen's Institute at Bristol. On March 11th a visit was paid to Uxbridge; and then no

lecture followed until the end of May, when " Pond
Life " was given at New Southgate, closely succeeded
by " The Bird," " A Country Walk," and " Life Underground " at Yarlet Hall, near Stafford.   Finally, on the
9th of July, the first part of " The Whale " was given
at Dulwich College ; and the second·part of the same
lecture, delivered two days later, brought the disastrous
season of 1884-85 to a close.

Only forty-four lectures had been delivered in all,
or little more than a third of the average annual number;
and the proceeds even of these had been almost entirely
swallowed up by the heavy expenses attending the
American tour.   No wonder my father made up his
mind never again to undertake another transatlantic
journey, although he had been warmly and repeatedly
pressed to do so before leaving Boston.   One such
experience was sufficient for a life-time, and he sted-
fastly adhered to his resolution, although subsequent
overtures were more than once made with the
view of inducing him to break it.   Lecturing in
England was more certain, more satisfactory, and
attended by far less of personal inconvenience.   The
fee paid for each individual lecture might not be so
high, no doubt ; but this difference was practically
equalised by the saving in travelling expenses and the
cost of living, which latter is in America at least one-
third more than in England.   And, besides, there was the
comfortable knowledge that, even if a general parlia-
mentary election were to take place, the British public
would not in consequence be rendered utterly and

totally incapable of talking or thinking of any other subject for at least three months to follow.

And so all idea of returning to America for a third lecturing tour was given up at once and for ever. A fresh syllabus was prepared, with details of additional lectures, and arrangements were quickly made with the view of making the English season of 1885-86 as successful as possible, and of regaining, so far as might be practicable, the ground which had been lost during the two preceding seasons.

# CHAPTER XVI.

## THE SKETCH-LECTURES (*concluded*).

THE season of 1885-86 began rather late with a lecture at New Southgate, on October 6th; and it soon became plainly evident that the two seasons of absence in America had done their work. For the continuity of the English lecturing had been broken; engagements without number had necessarily been refused; other lecturers in the same line had come to the front: and, of course, the lost ground could not be at once regained. Indeed, I do not think that the lecturing was ever again quite what it had been before. The interest and success of the lectures themselves were as great as ever, but engagements came in less frequently, and there was never again a season like that of 1881-82, when more than one hundred and twenty lectures were delivered in all. And it is difficult to attribute this falling off to any other source than the prolonged absence from the

English platform, which occurred at an unfortunate time, and was an undoubted error of judgment.

During the first three months of this season scarcely any lectures at all were given, and nothing in the way of an extended tour was possible. Early in the new year, however, a visit was paid to Ireland, and there a short series of lectures was delivered in several of the principal towns, concluding with the old favourite, " Unappreciated Insects," at Armagh.

This, by the way, was by no means identical with the same lecture as originally given. My father seldom gave any lecture twice upon precisely the same lines, but would add here, and prune there, as the whim of the moment prompted him, or as he saw that the attention of the audience was specially taken by any particular point. And also, whenever he was at home for any time, he made a rule of carefully revising, or even re-writing, his notes, adding fresh drawings, or improving those which he had, and so importing a constant element of novelty into each lecture, and preventing it from becoming monotonous and wearisome to himself.

And these changes had been, perhaps, more numerous in the " Unappreciated Insects " than in any other lecture. He began with quite a long list of insects, and reduced them at last, I think, to three—the Cockroach, the Earwig, and the Gnat—finding so much to say about each that very often the stipulated hour and a quarter was increased by half as much again, and the subject even then only partially treated. On more than one occasion, in fact, he confined his remarks to the cockroach alone, and delivered an entire lecture upon it. And so, also, in several other cases a lecture subject was gradually so enlarged that it had at last to be discussed in two parts as two distinct lectures.

During the latter half of this season lectures were a little more numerous than in the earlier portion ; but, nevertheless, only fifty lectures were delivered in all, as against more than double that number in preceding sessions. Yet the season, though less profitable than it might have been, was not quite so barren as it appeared. For, in other years, a number of lectures had been included, which, being delivered at neighbouring schools at such dates and hours as happened to suit my father's own convenience, were not remunerated very highly, and were, in some cases, given for little more than a merely nominal fee. After we left Norwood for St. Peter's these lectures were, of necessity, discontinued, and so, although the total number of engagements fell considerably, the receipts did not decrease in exactly the same proportion.

The following season began at Exeter, on September 30th ; and, a few days later, began a short but busy tour in the West of England. This included a special lecture, given at the Seamen's Institute at Bristol—an entertainment which proved so gratifying to the large and chiefly sea-faring audience that they showed their appreciation by singing

"For he's a jolly good fellow"

as the lecturer left the platform. At Windsor, where a

lecture was soon afterwards given, so many auditors
were attracted that the room was filled to overflowing,
and a number of late comers had to be accommodated
on the platform; an arrangement which proved highly
satisfactory to themselves, but highly annoying to the
lecturer, who objected even to the presence of a chair-
man upon the platform, and always liked to have as
much space as possible to himself. And the first
part of the season concluded with a flying visit to
Ireland, followed by an equally brief expedition to
Scotland.

While at Armagh, on December 12th, an amusing
incident took place, in the form of a little friction with
the police, who entertained grave suspicions as to the
true purpose of the type-writer, which now invariably
accompanied my father upon his travels. One of the
periodical dynamite scares had just been running its
course; detectives were everywhere on the watch for
suspicious characters and suspicious packages; and one
of these enlightened individuals persisted in believing
that the instrument in question was a new kind of infernal machine! The little bell, especially, which gave
notice as the end of each line was approached, appeared
to rouse the direst apprehension in his breast. And
nothing would do but that the type-writer must be
examined by the proper authorities before my father
could be allowed to proceed. Of course, its real nature
was quickly demonstrated, and he was allowed to drive
off with it in triumph; but the zealous detective evidently considered that he had greatly distinguished

R

himself, and had shown much *acumen* in arresting so distinctly suspicious an individual!

Lecturing began again very soon after Christmas, and the greater part of January was taken up by a tour in the northern counties. February was a fairly busy month, and March would have been so likewise, but for an unfortunate slip upon a slide at Upper Norwood. This resulted in an injury to the smaller bone of the right leg, which was cracked, although not actually fractured, and which for a time gave some little trouble. In spite of the damage, however, my father continued his tour for a few days, but, erysipelas supervening, was compelled to relinquish his remaining engagements, and return home for rest and nursing. In a few weeks, however, he had recovered sufficiently to deliver a short course of four lectures at Newbury, the population of which, from being entirely apathetic, had now become his warm admirers; and these brought the sixty-two lectures of the session to a close.

The following season—that of 1887-88—was of almost precisely similar character, although October and November were especially busy months. Work began on October 3rd, with a third visit to Ireland, Armagh, Coleraine, and Belfast, being the towns visited; and sixteen lectures in all were delivered during the month. Exactly the same number were given in November; but, on the other hand, the December engagements were only four in number, and those of January, 1888, but five. Seven lectures were given in February, six in March, and nine in the four following months; the season thus again including sixty-two lectures.

October 11th, 1888, saw the beginning of my father's last lecturing season. On that day he visited Malvern, and gave two lectures, the first on "Insect Transformations" at three o'clock, and the second, on "The Horse and his Master" at eight. Both this month and the next were fully occupied; and by Christmas he had delivered thirty-two lectures in all.

His life's work was now very nearly at an end. I returned home towards the end of November, for a visit of a few weeks, and noticed little change in him, except that he seemed rather more absent than usual, and rather more susceptible to fatigue. But he was away lecturing during a considerable portion of my stay, and, when at home, was generally hard at work in his study, after his manner; so that I saw comparatively little of him. On the 8th of January he started off to spend a week with some friends at South Norwood, and on the morning of that day I saw him for the last time.

I cannot help thinking that even now, although no traces of the disease which carried him off had as yet appeared, his constitution was breaking up. Now and then—as I afterwards learned—he would give way to drowsiness, and would fall asleep in his chair, or even at the table. He took cold repeatedly, and complained more than usual of pains in the chest. And the strain of the incessant work and the constant railway travelling seemed to be telling upon him, although at times

R 2

he was as bright and cheerful as ever. But yet he
never stopped his writing for a single day, and no
falling-off was noticed in the interest or the humour of
his lectures.

On January 15th, 1889, lecturing began again with
" Ant Life of the Tropics," at the Congregational Schoolroom, Ramsgate, where he constantly lectured, and was
always warmly welcomed. Other lectures followed, the
most noticeable being that on " Ant Life " at the London
Institution, which met with such marked success that
the committee, with a highly appreciative letter, sent
him an increased fee. This was on February 18th. On
the 19th he gave "Nature's Teachings" at New
Southgate; and on the following day he travelled north
for a short tour in Scotland.

To this, strange to say, he had from the very first
looked forward with absolute dread. Previous Scottish
tours had proved very successful, both from an artistic and
a financial point of view, and he had made many friends
who were glad again to receive him with their usual
hospitality. But he had always complained bitterly of
the terribly slow travelling upon the branch railways,
the long waits at the numerous junctions, and the
miserably inadequate accommodation provided for
passengers at all but the principal stations. An entire
day would sometimes be occupied in travelling fifty or
sixty miles; he would arrive at his destination with but
barely sufficient time to put up his frame, and without
a moment for rest or refreshment. And he had so vivid
a recollection of the discomforts which he had endured

on previous occasions, that, from the very time that this present tour was arranged, he repeatedly expressed a wish that he could avoid it. Only a few days before leaving home for the last time, indeed, he said that if there had been one lecture less arranged, he would even then, at the eleventh hour, relinquish the tour altogether, rather than go through what he knew lay before him. The shadow of what was to come seemed upon him, and he never left home for a lecturing expedition in such marked depression of spirits.

However, engagements were engagements, and he had never yet disappointed the public through any fault of his own ; and he accordingly left London for Glasgow on February 20th, and lectured at Tillicoultry on the next day. On the 22nd he went to Dollar, where he lectured in the evening. The next three days he spent at Fettes College, near Edinburgh, lecturing twice, and remaining over the Sunday ; and on that day he made the last entry in his diary—a mere brief reference to some letters which he had written.

On Tuesday, the 25th, he travelled southwards, and on the following day was to have lectured at Sedbergh Grammar School, in Yorkshire. But drowsiness overcame him in the train, he passed the junction at which it was necessary for him to change, and was unable to return in time to reach the school.

On this day the illness which carried him off assumed its final form. Even on Sunday, the 17th, he had been very unwell, and by advice had stayed quietly indoors until the evening, when he had promised to

preach for a clerical friend. During the week he took
cold, and failed to throw it off as usual. And, finally,
while waiting upon the platform of one of the Scotch
railway stations on a bitterly cold and windy day, he
had sustained a severe chill, which a few hours later,
under similar circumstances, was repeated. This led to
internal inflammation. Still he was not in the least
aware of his real state. Although suffering much pain,
he refused to see a local physician—by doing which, as
it afterwards appeared, he might have saved his life—
and merely wrote to his own medical man, describing
his symptoms, and inquiring whether or not they were
serious. In his home letters he did not mention the
subject of his health ; and still he persisted in going on
with his work.

On the night of Wednesday, the 26th, he did not
sleep at all, and on the Thursday was much weakened
by pain and want of rest. Still, however, he travelled
to Burton-on-Trent, and lectured on "Pond Life," at
St. George's Hall, in the evening. On that night again
he did not sleep, but yet contrived on the Friday to
revise a number of proof-sheets for the press, to write a
short letter home—still without any reference to himself—and to lecture again, for the last time, in the
evening. But he did so only with the utmost difficulty.
He could scarcely pull himself together to deliver his
lecture ; he was compelled more than once to leave the
hall, in order to obtain a little warmth at the fire in the
ante-room; and finally he brought his remarks to a
close rather suddenly, omitting the carefully elaborated

peroration with which he usually concluded. Yet those who were present said that the lecture was as interesting as ever, and the drawings as rapid and exact, although the lecturer was obviously suffering much pain, and clearly unfit to be lecturing at all. After the lecture the drawing-frame was taken down and packed as usual, and my father went to the house of some friends for the night.

Here his appearance excited so much alarm that he was entreated to see a doctor at once, and to relinquish all idea of fulfilling his remaining engagements ; but he still failed to realise his own condition, and insisted on starting early next morning for Coventry, where he was to lecture on the following Monday. During the night he became weaker, and there seemed a latent meaning in the words with which he thanked his host for certain arrangements made for his comfort during the journey— " *You know I would have done as much for you.*" But he took out his type-writer in the train, and attempted to write an article, of which, however, he only completed about half-a-dozen lines.

Shortly before five o'clock in the afternoon he arrived at Coventry, and drove at once from the railway station to St. Mary's Hall, where the lecture was to be delivered. There he left his apparatus, saying that he would come on the Monday afternoon to put up the screen. Thence he proceeded to the house of his old friend, Mrs. Bray, with whom he had promised to stay until the Tuesday morning, and who had been principally instrumental in arranging for his previous lectures in the town.

So far his wonderful courage and determination carried him on ; but now, at last, he was compelled to give way. Almost immediately on arriving he said that he feared he was very ill, and, after resting for a few minutes, asked that a prescription which he had with him might at once be made up by the nearest chemist. Alarmed by his appearance, however, Mrs. Bray suggested that a doctor should be immediately summoned, a proposal to which he at once assented ; and in a very short time one of the leading physicians of the town arrived, only to find him in a state of utter collapse from the combined effects of pain, exhaustion, and want of sleep. A brief examination sufficed to show that the patient was in a most critical condition, and suffering from an attack of acute peritonitis, with very serious complications ; so serious, indeed, that recovery, humanly speaking, was impossible. Of this my father seemed to be fully aware, for when, in answer to his own earnest inquiry, the physician told him the exact state of the case, he merely replied, " Just as I expected," and asked how many hours he had to live. Being told that in all probability the end would come in the course of the following day, he received the information with perfect calmness, walked upstairs with a little assistance, and was put to bed. By this time the pain had become so intense that injections of morphia were administered, as the only means of keeping the agony at bay. Under the influence of these he became easier, and slept during a great part of the night and the Sunday morning.

Several times he was asked whether a telegraphic summons should not be sent to my mother; but to this proposal he steadily refused his assent, on the ground that her own state of health rendered her unfit for the travel and anxiety. Neither would he allow the truth to be told her, for fear of the shock which it would cause. All that he would permit was a simple preparatory message, stating the fact of his illness, without details of any kind. And then he set himself to prepare for the end.

Early in the Sunday afternoon the pain left him—a sure sign that he had not many hours to live. For a time he seemed to rally, and, asking for pencil and paper, spent the next two hours in writing a letter home. Save that this letter is written in the faintest of characters—probably owing to the hardness of the pencil—no one would imagine for a moment that it had proceeded from the hand of a dying man. The writing is as firm and steady as usual, there is not a trace of incoherence, or even of haste, and, amid the many directions given, there is nothing superfluous, nothing that would have been discovered in the ordinary course of things without his assistance. His mind must have been perfectly clear and under control, in spite of his great weakness and the nearness of the end. And he was evidently fully aware of his condition, and quite conscious that those few pencil lines were the last that his hand would ever trace.

His letter written, he signed it with his usual firm, free signature, and asked that it might be posted

in time for the evening despatch.    And then for nearly
two hours he lay absorbed in prayer and meditation.

At six o'clock he complained of thirst, and asked for
a cup of milk.    Still his mind was perfectly clear, for,
finding that he could  no longer raise  his head to drink,
he asked whether there happened to be  an invalid's cup
in the house, and, finding that there was not, suggested
that a small milk-jug would answer the purpose instead.
This  was  procured,  and  he  drank  his  milk,  asking
immediately afterwards for a large cup of tea, which he
drank  also.    And  almost  immediately  afterwards  he
turned  his  head  upon  one  side,  and  quietly  passed
away.

# CHAPTER XVII.

## HOME LIFE.

AUTHORSHIP, although a sufficiently laborious profession
to those who depend upon the pen for their daily bread,
is commonly supposed to have this great counter-
balancing advantage, that it can be carried on quietly at
home, without any necessity for a daily journey to and
from an office, perhaps many miles away. This is no
doubt true enough; but it may yet be questioned
whether the ordinary man of business is not more
favoured than the author after all. For, although he
may be obliged to leave home early in the morning,
although he may not return until the shades of evening
have fallen, and although his work throughout the day
may be wearisome and exacting, yet, when he leaves

his office in the afternoon, that work is done. He has his evening to himself, to employ as he sees fit. He can enjoy a quiet pipe after dinner, or a friendly rubber, without feeling that he is neglecting any duty. He need not burn the midnight oil, unless it please him so to do. And he can hear the postman's knock with equanimity, for he has no dread of receiving a number of communications requiring an instant answer, or an urgent demand for MS. to be supplied without delay.

But with a successful author matters are very different. He may, perhaps, if he be methodical—as very few authors are—so map out his day as to devote certain specified hours to writing, in order that the rest of his time may be free. But this system of arrangement, if it be carried out at all, generally proves more theoretical than practical. A heavy batch of proof-sheets arrives unexpectedly, and must be revised without delay. An index has to be made out; an order arrives for an immediate article; or a number of books have to be reviewed at a few hours' notice. Probably a printer's emissary puts in an appearance, with orders not to take his departure until supplied with further "copy." And then the "system" has to be broken through, and the programme cast to the winds.

And in the case of an unmethodical writer, matters, of course, are far worse. He is never his own master. There is always work waiting to be done which should have been done days before, and which must now be done at high pressure if engagements are to be respected and kept. He may, perhaps, work hard all day, and yet there is a good deal to be done at night, quite irrespectively of any special orders which may come in by the evening post. And, consequently, although he may live at home, and work at home, he yet enjoys very little of what is generally termed "home life." All his time is spent in the study, and the more he does the more there seems to do.

Such was the case with my father throughout almost all his career. He could scarcely, perhaps, be described as unmethodical, for he was never behindhand with his work, and prided himself on always sending in his MS. rather before the time appointed. But he never would, and never could, tie himself down to write at certain regular hours; and, as he nearly always had an immense amount of work on his hands, the consequence was that frequently we seldom saw him, except at meal times, for days and weeks together. During the last few years of his life, indeed, strangers saw far more of him than ourselves, for after his lectures were over, and often upon intervening days, he was of course obliged to yield in some measure to the demands of society, and to spend in recreation hours which at home he would certainly have devoted to work. And no doubt this in great measure tended to the preservation of his health.

For, when at home, he was terribly careless of himself. He would work, day after day, with little or no intermission, taking scarcely any exercise, and sometimes not leaving the house for a week at a time. If visitors came he seldom saw them. All his time was passed in work. And probably only the enforced periods of comparative leisure, which now and then occurred in his lecturing tours, saved him from a complete break-down.

His power of work was simply astonishing. When I first remember him he was always at his desk by half-past four or five o'clock in the morning at all seasons of the year, lighting his own fire in the winter, and then writing steadily until eight. Then, in all weathers, he would start off for a sharp run of three miles over a stretch of particularly hilly country, winding up with a tolerably steep ascent of nearly a quarter of a mile, and priding himself on completing the distance from start to finish without stopping, or even slackening his pace. Then came a cold bath, followed by breakfast, during which his attention was entirely engrossed by his letters and the newspaper. After this an hour or so would be occupied with correspondence, which generally included the answering of natural history questions from all parts of the world. For all sorts of people used to write to him on all sorts of subjects, and though these manifold queries occasioned great demands upon his time and patience, and were often of the most trivial and even ridiculous character, he never allowed them to remain unanswered, but always sent at least a few courteous lines in reply. Autograph hunters, too, troubled him a good deal; but to these he replied much more curtly, generally returning the letter of request with the addition of the simple words, " Here it is, J. G. Wood."

By way of a specimen of his correspondence with

intimate friends I quote the following letter, which was written to a cousin shortly after the delivery of a special sermon in the early part of the year 1887. It is, I think, a fair example of his letters, and is further interesting as giving his own description of his style of preaching :—

> I am very much pleased to hear that your people appreciated my small efforts. As Mr. Swiveller remarks, after punching Quilp's head, "There's plenty more of it at the same shop; a large and extensive assortment always on hand. Country orders executed with neatness and despatch." So, if my good fortune should take me to C——, I shall be happy to place at your command the contents of my very limited treasury. I do not go in for rhetoric or special doctrines, and limit myself to the safe grounds of general exegesis.
>
> I very much enjoyed my L—— visit, and very much regret that M—— P——, like J——, has laid herself up by thoughtless imprudence. Your charming sex is an awful handful to manage, and whenever the judicious husbands are out of the way, the wives, like young sticklebacks, indulge in forbidden vagaries.
>
> The enclosed stamps are all that I can find at present, but if E—— should find them useful, they are very much at his service, and, for the future, I will preserve all that come to me. I hope that G—— may at last soften the Archbishop's heart. But Bishops, and, much more, Archbishops, seem to withdraw themselves into their own Olympus above the clouds.

Correspondence completed, another hour or so would be devoted to the revision of proof-sheets for the press —always a labour of time with my father, for he made many additions and alterations, and generally returned the sheets to press with manifold paper strips gummed all along the edges, each containing two or three lines of additional or substituted matter. Then he would go back to his writing until luncheon at half-past one.

And by this time he had usually accomplished a tale of work which with most writers would hardly have been completed in the entire day.

But his day's labour was by no means concluded yet. After luncheon, in those days, he always lay down until half-past three or four upon a couch in his study; but then he went back again to his desk until dinner at seven, and ordinarily again for a couple of hours more before retiring to rest. So that fully twelve hours out of the twenty-four, as a general rule, were spent with pen in hand, recreation being reduced to a minimum, and indeed almost to the vanishing point.

At this time there was certainly some little amount of regularity in his programme; but in after years, when a greater amount of sleep became an absolute necessity, he used to get through his work by the simple expedient of entering his study the first thing in the morning—generally about half-past seven—and scarcely leaving it, except for meals, until half-past eleven or twelve at night; and during almost the whole of the intervening time he would be working in one way. or another. He might write all one morning, and never touch a pen the next. But then he would be repairing a drawing-frame, or practising his great coloured sketches, or making out careful and elaborate notes for book, article, or lecture. And even at meal-times— dinner alone excepted—he was invariably engrossed with a book. He read at all sorts of odd moments: while dressing, or putting on his boots, in the train, or out walking. If he paid a call, and happened to be kept waiting for a few minutes, he was sure to be found deeply interested in some book taken up at random from the table. Yet he assimilated and remembered all this seemingly desultory reading in the most marvellous manner. He could usually quote all the most striking passages from any book or poem that he had read more than once, and kept all the vast amount of information that he had picked up on various subjects pigeon-holed away in the recesses of his own mind, yet so arranged and labelled, so to speak, that he could lay his hand upon it at an instant's notice. And he also possessed the great faculty of intuitively selecting the few items of value from the worthless mass, so that the latter passed away at once, while the former remained with him for ever.

These heavy and incessant labours my father maintained, with scarcely a day's intermission, to the end of his life. Indeed, in all my memory of him, I cannot recollect that he more than once gave himself a holiday of above a week's duration. And even his brief occasional periods of leisure were usually occupied in visiting museums, menageries, or libraries, and making out copious notes for future use. He had a perfect multiplicity of note-books, of all shapes, sorts, and sizes, from a microscopical volume of scarcely more than postage-stamp size, which he carried in his waist-coat pocket, to the bulky quarto in which were arranged drawings, photographs, cuttings, and extracts on almost every subject under the sun. For he by no means confined himself to the topics of his own professional

s

labours, but was well-informed on nearly every branch of science, art, and literature, and took almost an equal interest in all. Two huge extract-books now in my possession are filled with curious scraps of information on almost every imaginable subject, interspersed with portraits of living or recently deceased celebrities, short poems, generally of a humorous character (for he was extremely fond of comic poetry, if characterised by true wit), and biographical notices extracted from the daily newspapers. Besides these he had fifteen or twenty pocket-books, each given up to some special purpose, three or four diaries in simultaneous use, a "memoriser," in which were jotted down engagements—principally those for his sketch-lectures—and a russian-leather case or two for the safe carriage of letters or cuttings. Then all his books, and especially those of reference, were fully and carefully annotated. In his working-copy of his own Natural History there is nearly as much information in the form of fly-notes, pen-and-ink additions, and neatly inserted extracts and sketches, as in the three bulky volumes themselves. And, finally, for every book which he wrote he constructed a special note-book, in which were jotted down brief references for every branch and detail of the ground which the work was intended to cover.

These notes, however, were usually very meagre, and served merely to quicken his own memory as to things which he had seen or read. A key-word or two, and perhaps a few statistics : that, in many cases, is all ; and the record, to anyone but himself, was practically unintelligible. Yet in his own mind he carried the master-key to all this seeming disorder. He always knew what information he had upon any given subject, and in which of his multitudinous note-books to look for it. So that he never wasted any time in reducing this seemingly hopeless chaos of confusion to order and arrangement.

So, too, with his own study. To anyone but himself it presented a scene of utter disorder. A pile of books lay heaped untidily around his chair; for, as he finished reading, he always put his book down upon the floor beside him, so that a gradually increasing pile surrounded him until he had one of his occasional fits of order. Bones, skins, horns, curiosities of all kinds lay scattered about the floor. His work-table was covered with books, papers, letters, pamphlets, and a perfect infinity of miscellaneous objects, large and small. On another table near by was a second collection, equally varied, and even more extensive. On a third stood a perfect stack of magazines and periodicals of all kinds, almost entirely consisting of those in which his own articles had appeared. And numbers more stood upon the shelves, numbers more were piled underneath the couch, and numbers more still carefully tied together, and put away in boxes.

But this was by no means all. In different parts of the room were cages or boxes containing creatures specially under examination. There might be some scorpions in one, half a dozen snails in another, and a hedgehog or a blind-worm in a third, besides a number

s 2

of aquatic insects in a large bell-shaped vessel of water. Possibly in a shallow jar upon the table, filled with water, and with a layer of bees'-wax at the bottom, would be a careful insect-dissection. The microscope hard by would be ready for instant use. And then— scattered about in odd corners, on the shelves, under the shelves, even on the couch itself—would be a heterogeneous mixture of odds and ends of all descriptions, each in reality having a place of its own, but all in appearance constituting a scene of the most admired disorder. We always used to say, indeed, that my father's study was more like a marine store than a room in a respectable house; but he was quite unmoved by such sarcasms, and nothing annoyed him more than to have any object in his room interfered with.

And if any object *were* removed, he had a most wonderful way of discovering its absence. In this respect I fear I was myself a very frequent offender, for I used to borrow books or tools in his absence for some temporary purpose, and almost invariably forgot to replace them. After a time he found this out, and, whenever anything was missing, used to go straight from his room to mine to institute a search; in which, I am bound to say, he was usually successful.

His own writing-table was a kind of small exhibition in itself, for he always liked to have ready to his hand such articles as were in constant use, and made provision accordingly. So a small leather cross-piece on one leg of the table held a pair of enormous scissors, a band higher up held a smaller pair, while in a small

box screwed down to the table itself were a finer pair
still, used almost entirely for dissecting purposes. In a
number of holes bored through the woodwork at the
back were an assortment of bradawls and gimlets; a
small pocket just beneath the desk held a paper-knife;
one a little lower contained a two-foot rule; close
beside this was fastened a pin-cushion. From a tack on
one side hung a pen-wiper, while from a similar tack
upon the other depended an almanack. A little farther
on was a wire basket for holding letters which required
an answer, while from a nail close by was slung a large
slate, with sponge and pencil attached. And these
were so arranged that one and all could be reached
without the necessity for rising from his seat. On the
wall close by hung a series of hooks, one for revised
proofs, one for receipts, one for answered correspondence,
and so on. And a larger wire basket, within easy
reach, contained time-tables, a local directory, and one
or two similar chronicles in almost constant use.

Tolerably accomplished at most arts requiring skill
of hand, my father was a capital amateur carpenter, and
was perfectly competent to undertake most of the little
matters for which a professional workman is usually
called in. Of his tools he was exceedingly careful, and
kept the greater number under lock and key—chiefly, I
think, to prevent me from "borrowing" them. He
was also a very good locksmith and bell-hanger, and
even a plumber in a small way; and only in very exceptional cases was it necessary to call in a workman
while he happened to be at home.

Upon one point he was always extremely fond of
insisting, and that was the exceeding usefulness of oil.
Once a fortnight or so he always promenaded the house,
oil-bottle and feather in hand; and never was there a
door-hinge that squeaked, or a lock that refused to turn,
while he was in the house.  Every ball-cock was carefully lubricated; every screw was oiled before it was
driven home.  And thus, no doubt, he saved himself
and others a vast amount of subsequent trouble, on the
principle of the " stitch in time."

Another compound of which he was remarkably
fond was " sealing-wax varnish."  This he manufactured for· himself, pounding up a stick or two of
scarlet wax into a fine powder, dissolving it in spirit of
wine, and so producing a brilliant scarlet paint, which
dried quickly, did not crack, and was not easily rubbed
off.  Of this he made great use.  All his travelling
trunks and bags had his name, or his characteristic
monogram, painted upon them—in sealing-wax varnish.
The cases of his drawing-frames were encircled by
three broad rings—of sealing-wax varnish.  Even his
umbrella-handle was curiously marked—with sealing-wax
varnish.  And every six weeks or so he would give
up an hour to renovating these markings, performing
the operation with great care, and no little attention
to detail.

He was really a first-rate amateur bookbinder, again,
and mostly had three or four volumes under restoration.
He even tried soldering at one time, but I do not think
that he was very successful at this, as we never heard

very much about it, and he was distinctly reticent when the subject was mentioned. And he had quite a won-derful talent for inventing in a small way, and managed to make a lot of curious contrivances of no little in-genuity and great practical value.

Owing to the severe accident to his right hand already referred to, my father towards the end of his life was visited with threatenings of the dreaded "writer's cramp." Since the time of the accident itself —or rather since that of the comparative recovery which was all he ever enjoyed—he had never been able quite to trust the hand, which was frequently visited with nervous tremors and twitches, obliging him to steady it with the left when writing, or holding a cup or a tumbler. But when the more pronounced symp-toms set in, and grew daily more decided, he procured a portable type-writer, and thenceforward discarded the use of the pen altogether, except for correspondence, and sent printed instead of written MS. to the printers.

For this alteration, no doubt, the compositors were truly thankful, for, while his hand-writing in letters was beautifully neat and legible, his MS. hand was little more than a straggling series of microscopical hiero-glyphics, the difficulty of deciphering which was in-creased ten-fold by his inveterate habit of making numerous subsequent additions on small slips of paper, from every one of which, when affixed to the side of the sheet, ran a long line to the particular spot where he desired it to be inserted. As, in a single sheet of manuscript, there would sometimes be ten or a dozen of these additions, the lines from which crossed one another something like the foundation-threads of a spider's web; and as, moreover, corrections, alterations, and interlineations would be scattered about in profusion: the state of the MS. by the time it left his hands for press may be better imagined than described.

Yet, strange to say, the printers made few mistakes in setting up his work. Whether they put on a special man for the work of interpretation, or whether the compositors at last grew so accustomed to his writing that they found no difficulty in it, I do not know. But certain it is that they rarely misread it, and very often his proof-sheets went back to press with scarcely a correction marked upon them.

Of his proficiency with the type-writer he was very proud, and was always ready to show and explain the machine to any friend who might come to visit him After a while he always carried it with him upon the long railway journeys which his lecturing tours constantly involved. Indeed, during the last four winters of his life, almost the whole of his literary work was performed while actually in the train. From the very first he was perfectly careless of appearances, and utterly indifferent to any attention or excitement which his proceedings might arouse. He once walked through the streets of London carrying a collection of savage weapons, and on another occasion wheeled a barrow-load of bricks from the builder's yard to his house, as he wanted them for immediate use, and no workman happened to be at liberty. So that he was naturally quite unaffected by the interest which his type-writing proceedings excited among his fellow-passengers, who gazed their fill at the elderly clergyman in the corner of the carriage, busily performing upon the keys of the strange machine, and even made their comments thereupon in no inaudible tone of voice, without in the least arousing him from his abstraction, or impeding his flow of thought.

Yet, by a strange contradiction, when writing quietly at home in his own study, he was almost nervously sensitive to any interruption. The passing of a footstep near his door, a false note upon the piano downstairs, or the barking of a distant dog, would often upset him altogether, and render him unable to write for a quarter of an hour or more after the exciting cause had passed away. So that this perfect ability to write in the train was most remarkable. Well it was that he was able to do so, for his lecturing tours during the winter months were so long and so frequent that otherwise literary work must almost wholly have been put on one side during fully one-half of the year.

Strangely enough, neither bodily fatigue nor actual suffering (unless, of course, of a severe character) seemed noticeably to impair his power of writing. At the end of a long day's labour his style would be as pleasant, his ideas as free and fresh as if he were but just beginning; and although throughout his life he suffered greatly from dyspepsia, it had little or no effect upon his work. Probably the incessant stooping over his desk (just after meals especially) was largely responsible

for this last; and he did not improve matters by drinking large quantities of tea and coffee, of which he was
almost inordinately fond, and would partake from the
first thing in the morning until the last thing at night.
He was also subject to that curious but common delusion which holds that the efficacy of physic is in exact
proportion to the amount taken. If a small dose did
him good he would double it, and expect the beneficial
effects to be correspondingly increased. And in this
matter experience never seemed to teach him wisdom.
But his digestive system was weak from a boy; and,
though always simple and abstemious in his diet, he
was seldom altogether free from his constant enemy.

In other ways his health was wonderfully good. In
all the years that I can remember him I do not recollect
that he was ever really ill, save and except during the
anxious winter of 1877–8; and even such minor ailments
as colds were almost wholly unknown to him. No
doubt this immunity from sickness was due in great
measure to his early training as an athlete, which
strengthened every part of his frame, and transformed
him from a weak and puny boy into almost a model of
physical and muscular strength. A long and systematic
course of Turkish baths, too, rendered him almost proof
against cold, and also greatly improved his general
health. And so he was able, year after year, to work
day and night as he did, taking but little exercise, and
obtaining but little sleep, and often subject to a mental
strain and anxiety which would soon have caused a
weaker man to break down altogether.

Yet, strangely enough, in spite of his wonderful constitution, he was always nervously anxious about himself, would magnify a trifling ailment into a serious disease, and work himself into a perfect fever of alarm over an indisposition of which others would have thought nothing. But this is a not uncommon failing with those who are seldom ill, and who seem to appreciate no gradations between perfect health and dangerous sickness.

# CHAPTER XVIII.

## PETS.

Pets of Childhood—Pet Snakes—"Apollo" the Bull-dog—His mischievous propensities—"Roughie" and "Bosco"—"Pret" and his successors—"Grip" the Raven—An overdose of Linen—A pet Chameleon—Blind-worms, old and young—Feeding them with slugs—Pet Toads—Various Lizards—Tortoises—A Bat and its diet—Cage-birds—Outdoor Pets—Daily Pensioners—Fat *versus* Bread-crumbs—The Scene on the Window-ledge—Impatient Sparrows—Tit-mice and how they were fed—Pet Lions and Tigers—How to get on friendly terms with a Lion—Adventure with a cross-grained Dog—Special fondness for Cats—A Second Mahomet.

THROUGHOUT almost the whole of his life—as was indeed only to be expected—my father was in the habit of keeping pets of almost every kind. Long before even he went to school he was constantly bringing home creatures which he had found in his country rambles, not at all for the mere purpose of making pets of them, after the manner of most boys with a natural fondness for animals, but that he might keep them for a while in captivity, and watch their habits at home. During his school-days, snakes seem to have been his special favourites; not vipers, of course, but the evil-smelling grass-snakes, which, however, learned in time to retain their horrible odour unless when touched by the hand of a stranger. And in after-life there were few available creatures which he had not for some time

had in captivity, besides a perfect host of the more
commonly domesticated dogs and cats.

The history of several of these is given in
" Petland," and " Petland Revisited." There is the
account of " Apollo " the bull-dog, as ugly and un-
prepossessing a dog as ever lived, in the eyes of the
unprejudiced beholder ; but a pleasant, good-tempered
beast withal, and one whose fidelity and obedience were
something marvellous to witness. My father had him
from the veriest puppy, and trained him carefully himself; and the dog well repaid the care which was
bestowed upon him. He would even give up a bone at
the word of command—a very unusual concession upon
the part of a bull-dog—and was as tractable a creature
with those whom he knew as anyone could wish. For
myself, when a child of about two years old, he had a
peculiar affection ; and very often we changed places, I
taking possession of the kennel, and he sitting between
me and the entrance and mounting guard over me. My
nurse, however, was not in the least afraid of him, but
simply used to box his ears and haul him away when I
had to be extracted. But no one who was not on
intimate terms with him ever dared to approach him
at all.

But "Apollo" was a somewhat expensive dog to keep,
for once he ate a big hole in some park-palings, because
they were too high for him to leap over with a stick in
his mouth ; and then, finding that when he seized his
master's property by the middle, and tried to pass
through the hole, he naturally failed, he set to work

once more and ate another big hole for the stick! And
on another occasion he followed my father into church
through the medium of a stained-glass window. And
by-and-by "Apollo" had to be given up.

Then there is the story of "Roughie": a stupid
beast, who was never of the smallest use to anybody,
and cared for nothing and no one in the world except
himself. "Bosco," a skye-terrier, is only casually
mentioned in "Petland Revisited." He was the last of
my father's own dogs, and lived with us from infancy to
old age. "Jock," who came to us after we went to live
at St. Peter's, and has been the chief subject of more
than one magazine article, was practically my property,
and only acknowledged my father as a sort of deputy
master.

Of cats, there were a long series, beginning with
"Pret," whose life and adventures occupy the first five
chapters of "Petland Revisited." After him we had
more than I could possibly reckon up (eighteen in the
house at one time, inclusive of kittens!), most of which
have served as the subjects of an article or two in one
of the magazines. Our two present cats, "Bunny" and
"Fluff," have had most of their deeds recorded in *The
Child's Pictorial.*

We once had a raven, who came to us in a hamper,
went by the suggestive name of "Grip," and had a big
cage and a long wire "run" made for him in the yard
outside the kitchen. He was a vicious brute, and never
evinced the smallest affection for anybody, impartially
pecking at us all, whenever we went near enough to his

run, and making himself generally objectionable. He was a crafty creature, too, and whenever he saw a chance of pecking anyone, would ostentatiously show himself at quite the other end of his run, and be apparently quite oblivious of his opportunity. But his wicked eye was upon his anticipated victim all the while, and, the moment that he saw that his movements were unwatched, he would sidle along the run, and then drive his beak with all his force against the legs of the unwary visitor. And to the trousered and knickerbockered half of humanity such a dig invariably meant a deep puncture, and a smarting wound which did not heal for several days.

"Grip" died of too much linen, a couple of towels having been blown from a neighbouring clothes-line upon his run, and promptly torn to rags and demolished before they could be rescued. The bird did it out of pure mischief, and only ate the torn strips because he knew that he was doing wrong, and took a fiendish delight in doing it. But his meal naturally disagreed with him, and he died. And none of us mourned over his grave.

My father also once made a pet of a chameleon, whose biography appears in "Petland." This creature —one which was curious and interesting in the last degree, but to which even the most enthusiastic of naturalists could scarcely feel any particular sensations of attachment—fell a victim at last to the jealousy of "Pret," who, like most pet animals, could not bear to see any attention lavished upon any living creature besides himself, and took an opportunity of slaying his supposed rival in the temporary absence of his master.

Once a blind-worm was brought home, and kept for some time, serving, indeed, as the model from which the description in the larger Natural History was taken. After a short time, nine little blind-worms unexpectedly made their appearance, and presented the strange phenomenon of steadily increasing in size, apparently without taking food of any kind at all. Yet they went through all the actions of slug-catching, just as though they were quite big blind-worms, instead of little creatures not much more than an inch in total length.

"When I introduced the slugs," writes my father, "the odd little reptiles acted just as their mother was doing, followed the slugs about with their heads, hovered over them, and made believe to eat them, and then were quietly walked over by their intended prey ; which, being nearly twice as big as themselves, proceeded on their course without paying the least regard to the tiny reptiles, whose bodies were not larger than ordinary knitting-needles, and easily glided over them, or put them to ignominious flight."

Of toads my father had many, from the time when— "a small six-year-old naturalist, with a magnifying-glass always open in one hand, and an empty pill-box in the other"—he "used to potter up and down the garden" in search of any natural history object that might present itself. And he wrote a special article about them in *Once a Week*, which afterwards was reprinted in *Out of Doors*. Lizards, of course, he had of various kinds, among them a specimen of that strange, spine-covered creature, the Tapayaxin, which was sent him *by post* from North America, and arrived in safety, only however to expire after a very few weeks in captivity. Newts he frequently kept, generally in a globe of water in his study, or in one of those large oblong glass tanks generally dignified by the title of "aquaria." Of tortoises there were several at different times, the last being fastened to a post in the garden by means of a long string passed through a hole in the edge of its shell ; and two chicken tortoises lived for a long time indoors. Then there came half a dozen scorpions from the south of Europe, which theoretically were mine, but practically were monopolised by my father; and two great *Bulimus* snails from the West Indies, which lived for some months under a glass globe without doing anything at all to justify their existence ; an injured bat, which was provided with a similar domicile, and had to be supplied with blue-bottle flies at the rate of about seventy *per diem*, or rather *per noctem ;* and mostly a few odds and ends in the shape of water-beetles, dragon-fly grubs, or some other creatures which happened at the time to be more particularly under observation.

For cage-birds my father never manifested any fondness, save that he once kept quite a large number of canaries for some time in one large cage, knew them all separately and by name, and took the greatest interest in their doings. And once we had a parrot:

T

a queer bird, which displayed quite an unusual aptitude for bringing out appropriate remarks, but which
had the temper of a demon. But, broadly speaking, my
father looked upon the imprisonment of birds in cages
as little short of positive cruelty, and often said that
they were so changed by prolonged captivity that they
quite ceased to be their own natural selves at all.

Besides all these, my father had quite a number of
what I may perhaps describe as *outdoor* pets. All
through the autumn and winter months, for instance,
he regularly fed a large number of birds at his study
window, mixing a large dishful of oatmeal porridge,
bread-crumbs, and scraps of meat, and placing the
contents on the window-ledge as soon as he went
upstairs after breakfast. Of the presence of meat—or
fat—in the mixture he always made a special point,
saying that warm-blooded creatures like birds needed
something more stimulating than mere bread-crumbs
and oatmeal to keep up the bodily heat, and that the
meat took the place of the worms and insects which in
weather less inclement they would have captured for
themselves.

The birds soon found out that this food supply was
part of the regular daily programme, and half an hour
or so before feeding-time they used always to assemble,
pushing and jostling one another in order to secure the
best places. The same birds came day after day, and
were soon well known by sight, thrushes, blackbirds,
titmice, finches, robins, and sparrows all contending
with one another for their share in the daily distribu-

tion. Of course the last-named got the best of it, as they always do, their natural pugnacity and aggressiveness bringing them well to the front, and enabling them to hold their own against seemingly overwhelming odds. And then would ensue a scene of rare confusion and quarrelling and scolding, each bird doing its best to secure the choicest morsels, and eating with all convenient rapidity in order to obtain as much of the enticing food as possible before the feast was despatched. But then, perhaps, the snow would be lying several inches deep upon the ground, and this morning refection would be almost the only food that they would be able to procure for the next twenty-four hours.

All these birds knew to the very minute when their regular feeding-time had arrived, and, if my father happened to be a little late, would tap impatiently and repeatedly at the window, in order to accelerate his movements. Sometimes quite a row of sparrows would be sitting there and pecking at the glass when he arrived with the dish. Then, of course, there would be a general flight as the window was opened, and the contents of the dish were spread out. But almost before it could be shut again every bird would be back, and the struggling for place and the strife of tongues would commence. This used to go on every morning until the warm days of spring brought the insects out; and then the attendance would gradually diminish until hardly a bird arrived at all.

The titmice were the most unfortunate of all these birds, for their small size enabled any and all of their

T 2

comrades to oust them from any vantage-ground which
they might have taken up; and this the said comrades
were never slow in doing. So my father put up near
his window a kind of permanent feast, to which the titmice, and the titmice alone, should have access. This
was managed by enclosing a number of small lumps of
suet in a little bag of large-meshed network, and sus-
pending it by three or four feet of string from an out-
stretching branch of a tall tree. The titmice, being
marvellously proficient in the art of climbing, and quite
as much at their ease when hanging head downwards as
when in the ordinary position, were of course perfectly
satisfied with this arrangement, and might be seen upon
the ball of suet at all hours of the day, pecking away
busily—not to say greedily—at the suet within, and
speedily reducing it to a mere tithe of its former dimensions. And very pretty indeed the odd little birds looked
as they clung to the network, swinging in the wind,
and eating their meal without fear of molestation from
the sparrows. For the latter, of course, were hopelessly
debarred by their physical structure from disputing
with them the possession of the dainty repast, and they
and their fellow-finches could only look on, and see
the titmice enjoying themselves, without the remotest
chance of becoming participators in the feast.

Other " outdoor " pets, too, he had at this time in
the shape of some lions and tigers at Margate. These
belonged to Sanger's menagerie, to which, as it was
within walking distance, he was a very frequent visitor.
Indeed, while he was at home, he very seldom let a week

pass without paying at least one visit to his favourite animals. And partly by his own natural talent for going wherever and doing whatever he wished, partly, I have reason to believe, by the still more potent aid of an appeal to the pocket, he soon contrived to gain admission, as it were, behind the scenes, struck up a great friendship with the head-keeper, and was permitted to do exactly what he pleased without any sort of interference from anyone. And he soon came to be on terms of the most perfect friendship with the lions and tigers, which would allow him to pull them about, examine their claws, and, in fact, do just whatever he liked with them. And all this amiability was due to his habit of taking a small bottle of lavender water with him whenever he visited the menagerie, sprinkling a few drops upon a rolled-up ball of paper, and then throwing it into the cage. The animals used to go nearly wild in their exuberant delight. They would grasp the ball with both fore-paws, hold it close to their nostrils, and then draw in a succession of deep inhalations of the fascinating perfume, purring loudly the while. Then would come, perhaps, a loud roar, expressive of deep enjoyment, and then more inhalations; and so on. And after he had visited them a few times they used to detect my father in the far distance, and dance about in their cages with excitement, roaring loudly, until he came up and produced his paper balls.

But my father had that peculiar knack, enjoyed only by a few, of making friends at once with any animal with which he happened to meet. No dog ever

failed to respond when he stooped to caress it; no cat
ever growled even when he took up her kitten and
stroked and admired it. All seemed instinctively at
once to recognise his love for them; and all showed
that they did so by the way in which they received his
advances.

Here is his own account of an adventure with a
peculiarly ferocious and cross-grained dog:—

There was a Scotch terrier dog who lately died, to the, very
great sorrow of his master, an officer in the 45th regiment, and the
very great rejoicing of his master's friends. He was good enough to
honour me by admitting me among his friends—the only person not
belonging to the family to whom he extended that privilege. His
name was "Mess," which was a military abbreviation of "Mesty,"
which was an abbreviation of Mephistopheles, the name being given
to him in consequence of his temper, which really deserved the name
of infernal. No one, except his master, his master's family, and an
exceptionally favoured servant or two, could put a hand on him
without being bitten. I know a learned barrister who has been kept
in bed until a very late hour in the morning, because "Mess" had
come into his room when the servant brought the hot water, and
would not allow him to get up. As long as he lay still in bed, Mess
sat quietly on the floor; but at the least movement, Mess sprang up
with a menacing growl, flashing eyes, and gleaming teeth, and the
unfortunate guest had to subside again, unable even to ring the bell
for help, and knowing that his host and hostess must be waiting
breakfast for him and chafing at his laziness.

One day I paid a visit to "Mess's" master, not knowing anything about the dog, and not seeing the dog when I arrived. Being
accustomed to an early walk before breakfast, I started off as usual
on the following morning, and, on returning, met a little procession,
consisting of a nursemaid leading a donkey, on which were the two
daughters of my host in panniers, and a remarkably fine Scotch
terrier, which was trotting along in front. As soon as he saw
me the dog sprang forward, and I, not knowing anything of his

character, and thinking that he wanted a game, stooped down, patted him, rolled him on his back, pretended to box his ears, put my hand into his mouth, and, in short, let him have his game. The nurse-maid stood by, almost paralysed with horror; but why she should be frightened seemed rather mysterious.

On coming to breakfast I spoke in high terms of the splendid dog with whom I had enjoyed a game, and the host was almost as horrified as the nurse had been. Not until then did I hear about the dog's temper; but, whatever it was, it was never displayed towards me, and I believe that I am the only person not belonging to the family who was ever allowed to put a hand on him.

While he was at meals, a favourite cat generally sat on my father's shoulders; very often there was one upon each. He was very fond of making them spring from the ground upon his shoulder, walk along his out-stretched arm, and take a piece of meat from between his finger and thumb; and I think that they took to sitting upon his shoulder merely that they might be nearer the scene of action, and so save time when the meat was held out. One of these same cats, also, had a special fondness for my father's own chair, which was cushioned in a manner which she particularly admired; and if he happened to go out of his study and leave the door open, he was nearly sure to find her there when he returned. Nor did she at all hesitate, if he turned her out and occupied the chair himself, to signify her desire for a share of his seat, for she would leap up, and deliberately push him with her paws until he moved. This would go on until he left her sufficient space to settle down comfortably. And there the two would remain for hours.

He would do almost anything for a cat. I have

known him to leave his work, and to go down from the top of the house to the bottom, on three separate occasions in the course of a single morning, just because it occurred to one of our cats that she would like to show him her kittens. She would come to the door, and mew for admission, and then sit and mew again until he got up and accompanied her downstairs; and then she would go straight off to the basket where her kittens were lying, and rub herself against it, and then against him, purring loudly, and in every way endeavouring to show her pride. Then he would stroke and admire them, ask the cat whether she were satisfied, and then go back to his work. And in the course of half an hour or so the same programme would be exactly repeated. I do not know whether he would have gone quite so far as the Prophet Mahomet is reported to have done, and have cut off the sleeve of a coat upon which a cat was lying asleep rather than disturb her. But he would always cheerfully give up his time to supplying the wants of a cat, or indeed of any animal, whether those wants were fancied or real. And he was never more happy than when surrounded by animals with which he was intimate, and which, to him, were not only companions, but true and actual friends.

# CHAPTER XIX.

### RECREATIONS.

Busy worker though my father was, and able to labour steadily on for weeks together with scarcely an hour's intermission, he was yet, by the very force of Nature, compelled to indulge in occasional periods of recreation ; and, during these, he contrived to make himself a thorough master of many games of skill. Mediocrity in these was absolutely unknown to him. He either played a game well, or he did not play it at all. But when he did play it, he studied it theoretically as well as practically, and sometimes devoted to it almost as much care and attention as though it had been a new branch of science ; while he lost no opportunity of observing the play of those more proficient than himself, in order that he might pick up a hint or two from their superior acquirements.

During his university career he became a most accomplished gymnast, and could perform with perfect ease numerous feats seldom attempted by an amateur. His course of practice was, in the first place, entered upon for prudential reasons. From his very birth he was weak and sickly, and during the years of his childhood, indeed, it always seemed doubtful whether or not he would ever attain to man's estate. But, by a long and careful course of training in the gymnasium and on the running-track, he contrived, not only entirely to overcome his constitutional weakness, but to build up the physique which alone enabled him in after-years to carry on his severe and incessant labours.

In his well-known "Adventures of Mr. Verdant Green," the late Rev. E. Bradley, better known as "Cuthbert Bede," has largely made use of my father, who stood for the character of "little Mr. Bouncer" in the chapters relating to the gymnasium. I quote the following from a short account of my father as a gymnast, lately contributed by Mr. Bradley to the *Boys' Own Paper*. Citing his own description in "Verdant Green," he says :—

"Opposite to the door was the vaulting-horse, on whose wooden back the gymnast sprang at a bound, and over which the tyro (with the aid of the spring-board) usually pitched himself headlong. Then, commencing at the further end, was a series of poles and ropes—the turning-poles, the hanging-poles, the rings, and the *trapeze*—on either or all of which the pupil could exercise himself; and, if he had the skill to do so, could jerk himself from one to the other, and finally hang himself upon the sloping ladder before the momentum of his spring had passed away."

Wood could do this work with apparent ease, and swing himself from one end of the room to the other, as though he were a performer in a circus. One of my sketches in "Verdant Green" shows him taking his aërial flight, and we used to compare him to one of the "Bounding Bricks of Babylon," so amusingly delineated by Keeley and Alfred Wigan in Albert Smith's burlesque extravaganza, *The Alhambra.* Wood fairly revelled in these gymnastic feats, which he performed with as much elegance as ease.

Substituting for his name that of Mr. Bouncer, I spoke of him as follows :—" Mr. Bouncer, who could do most things with his hands and feet, was a very distinguished pupil of Mr. MacLaren ; for the little gentleman was as active as a monkey, and—to quote his own remarkably figurative expression—was ' a great deal livelier than the Bug and Butterfly' (which was a name given to Mr. Hope's Entomological Museum). Mr. Bouncer, then, would go through the full series of gymnastic performances, and, finally, pull himself up the rounds of the ladder with the greatest apparent ease, much to the envy of Mr. Verdant Green, who, bathed in perspiration, and nearly dislocating every bone in his body, would vainly struggle (in attitudes like to those of ' the perspiring frog' of Count Smorltork) to imitate his mercurial friend, and would finally drop exhausted on the padded floor."

These gymnastic attainments my father kept up to some extent after leaving college, and, even after his marriage, he more than once ascended a tolerably high tree, in order to hang by his feet alone from one of the loftiest branches.

For cricket he did not much care. He played as a boy (although the game was then comparatively in its infancy), but met with a serious accident, breaking his right leg and dislocating his ankle through a severe fall. After this he seldom played, and, indeed, took no special interest in the game, beyond studying the-

reports of leading matches as given in the daily newspapers. For football I do not think that he ever cared
in the least. Lawn-tennis, of course, did not come into
fashion until his advancing years prevented him from
taking it up. Probably, the injury to his right hand
from the fall of 1874 would, in any case, have prevented
him from attaining to any proficiency in the game.
But he nevertheless took a very lively interest in it,
and would generally come out and watch us for a while
as we played in the afternoon.

Croquet, however, attracted him far more, and he
reduced it almost to the level of one of the exact
sciences. When we removed to the larger house at
Belvedere, he had the lawn carefully taken up and
re-laid, with a substratum of chalk, on the most approved principles. And then his custom was, when
tired with work, to come out and play a short game
with himself, blue balls against red. The game was a
very scientific one. Six hoops only, placed at sharp
angles to one another, at great distances apart, and no
more than five inches in width; heavy balls, carefully
selected and tested, that they might run as accurately
as a billiard-ball; and a large, heavily-loaded mallet,
with the handle carefully wrapped with string. These
were his requirements, and without them he did not
consider the game worth playing. Nothing annoyed
him more than to be compelled to take part in the
ordinary garden-party croquet. He always rebelled
against even the four-player game, holding that no real
opportunity for skill could be given unless the players

were restricted to two, with two balls each. But his misery when—as now and then happened—he was forced to serve as one of *eight* was really almost pitiable to witness. But at last he hit upon an expedient, which had the double advantage of quickly releasing him from bondage, and minimising the chance of his being again pressed to undergo a similar martyrdom; and, ignoring his partners altogether, he would simply run the round of the hoops at a single turn (which, with no less than seven balls to help him, he could easily do), and then "peg out" and go away. After a time this custom of his became notorious, the desired result followed, and the detested eight-ball game was no longer inflicted upon him.

Given what he considered as the proper requirements—the heavy balls, narrow hoops, loaded mallet, and lawn as true as a bowling-green—his play was wonderfully accurate, and he was especially good in his management of the balls with a view to their future use. Thus the next player would always find himself at the far end of the ground, with nothing to aim at which was not partially protected by a wire or a stake; while his own following ball was sure to be in position, with the fourth ball in readiness to help it on. The game, indeed, came with him to resemble billiards rather than croquet, and he played it with his head quite as much as with his hands.

Of billiards itself, too, he was very fond, and subscribed, while at Belvedere, to a small private club, so as to be able to play a game whenever he could afford

the time. Just now and then he used to play instead
of lying down in the afternoon—on which occasions I
usually accompanied him in order to act as scorer. But
the sleeplessness from which in those days he suffered
so much prevented him from giving up his nap unless
under exceptional circumstances ; and generally an odd
game or two in the course of the week was all that he
could manage.

Of swimming, in all its branches, he was a perfect
master, from his very earliest years. When barely
four years of age his father was accustomed to take him
down daily to bathe in the river Isis, and he thus
acquired a familiarity with the water which enabled him
to perform all the feats of the professionals with perfect ease. He could swim in a dozen different ways ;
dive to almost any depth, and from almost any height ;
swim under water ; and, in fact, perform in the water
pretty well all that it was possible for man to do.

And he was a very good long-distance swimmer,
thinking nothing of swimming half a mile out to sea
when taking his morning bathe. Once, upon one of
these excursions, he actually went to sleep in the water,
and remained quietly floating for some little time till a
sharp storm of hail awoke him. But on another occasion he met with a far more unpleasant experience,
which left its effects behind it for weeks, and barely
escaped being attended by the most serious conse-
quences ; for he swam into contact with one of the
terrible medusas, or stinging jelly-fish, whose envenomed
streamers passed across his breast, and injected their

deadly poison into a thousand tiny wounds. I quote the following from his own account of the accident:—

One morning towards the end of July, while swimming off the Margate coast, I saw at a distance something that looked like a patch of sand, occasionally visible, and occasionally covered, as it were, by the waves, which were then running high in consequence of a lengthened gale which had not long gone down. Knowing the coast pretty well, and thinking that no sand ought to be in such a locality, I swam towards the strange object, and had got within some eight or ten yards of it before finding that it was composed of animal substance. I naturally thought that it must be the refuse of some animal that had been thrown overboard, and swam away from it, not being anxious to come in contact with so unpleasant a substance.

While still approaching it I had noticed a slight tingling in the toes of the left foot, but as I invariably suffer from cramp in those regions while swimming, I took the " pins-and-needles " sensation for a symptom of the accustomed cramp, and thought nothing of it. As I swam on, however, the tingling extended further and further, and began to feel very much like the sting of a nettle. Suddenly the truth flashed across me, and I made for shore as fast as I could.

On turning round for that purpose, I raised my right arm out of the water, and found that dozens of slender and transparent threads were hanging from it, and evidently still attached to the medusa, now some forty or fifty feet away. The filaments were slight and delicate as those of a spider's web, but there the similitude ceased, for each was armed with a myriad poisoned darts that worked their way into the tissues, and affected the nervous system like the stings of wasps.

Before I reached shore the pain had become fearfully severe, and on quitting the cool waves it was absolute torture. Wherever one of the multitudinous threads had come in contact with the skin there appeared a light scarlet line, which on closer examination was resolvable into minute dots or pustules ; and the sensation was much as if each dot were charged with a red-hot needle gradually making its way through the nerves. The slightest touch of the clothes was agony, and as I had to walk more than two miles before reaching my lodgings, the sufferings which I endured may be better imagined than described.

Severe, however, as was the pain, it was the least part of the torture inflicted by these apparently insignificant weapons. Both the respiration and the action of the heart became affected, while at short intervals sharp pangs shot through the chest, as if a bullet had passed through heart and lungs, causing me to fall as if struck by a leaden missile. Then the pulsation of the heart would cease for a time that seemed an age, and then it would give six or seven leaps as if it would force its way through the chest. Then the lungs would refuse to act, and I stood gasping in vain for breath, as if the arm of a garrotter were round my neck. Then the sharp pangs would shoot through my chest, and so *da capo.*

After a journey lasting, so far as my feelings went, about two years, I got to my lodgings, and instinctively sought for the salad-oil flask. As always happens under such circumstances it was empty, and I had to wait while another could be purchased. A copious friction with the oil had a sensible effect in alleviating the suffering, though when I happened to catch a glimpse of my own face in the mirror I hardly knew it—all white, wrinkled, and shrivelled, with cold perspiration standing in large drops over the surface.

How much brandy was administered to me I almost fear to mention, excepting to say that within half an hour I drank as much alcohol as would have intoxicated me over and over again, and yet was no more affected by it than if it had been so much fair water. Several days elapsed before I could walk with any degree of comfort, and for more than three months afterwards the shooting pains would occasionally dart through the chest.

On two subsequent occasions this painful experience was repeated, but as in these cases the envenomed streamers touched the foot only, the effects were not nearly so severe.

On the ice my father was as accomplished as in the water. Like most who have thoroughly mastered the intricacies of figure-skating he cared for nothing else, and was perfectly satisfied with a small pond, so long as the ice was smooth and free from obstacles. And, given such a pond, he would spend an hour or so in perfect happiness, cutting threes and eights and all the various figures which only an accomplished skater can achieve.

He was also a fair long-distance runner, with considerable powers of endurance, but no great speed; and his daily three-mile run, which he kept up until about his forty-fifth year, scarcely seemed to fatigue him in the least. And at one time he was not unfrequently in the habit—on Sundays more especially—of paying a visit to a distant friend, travelling by train one way, and walking or running the other. Sometimes, when he reserved the train for the return journey, his adventures were rather amusing, for he never would allow himself more than just sufficient time to reach the station, and was not infrequently left behind. On one occasion he turned up at half-past three in the morning, having missed his train, but having managed to obtain a lift for part of the distance in a belated tradesman's cart. Another time he came home on the engine of a goods train, by means of a small *douceur* to the driver, who obligingly stopped just outside the station in order to allow him to descend. Of course he had no business to be there; but he had the most wonderful knack of going where he chose, and doing what he chose, without interference from anybody. He once obtained one of the very best stalls for the Handel Festival simply by asking for it; he found his way behind the curtain of a London theatre in order to

U

discover the secret of a thrilling "fire-scene," and no
one said him nay; he even in some mysterious manner
invaded the sacred precincts of the Speaker's Gallery in
the House of Commons, without the usual formality
of obtaining an order. And so well known was this
enviable faculty among us that we always used to say
that he would succeed even in getting presented at
Court if he wished, without any preliminaries or introductions whatever.

Of my father's *indoor* amusements, music was the
chief; and this, indeed, was almost a passion with him.
He was no vocalist, for his voice, good enough when a
boy, was irretrievably damaged by the common mistake
of continuing to use it after it had "broken," and,
though he could take his part in a chorus, it was quite
useless for solo work. Nor was he a pianist. The only
instrument, indeed, which he played at all was the
euphonium; admirable, perhaps, as a constituent in a
concert orchestra or a brass band, but scarcely suited to
the narrow limits of a drawing-room. No doubt his
performances must sometimes have been a sore trial to
the neighbours, more especially as in those days—when
I was about ten years old—I used to play the cornet
after a fashion, so that we might practise duets
together. We both belonged for some time, also, to
an amateur brass band at Erith, which met for a weekly
practice under the leadership of a band-master from
Woolwich. But my father's chief reason for taking up
his noisy but somewhat unwieldy instrument, and also
for inducing me to take up mine, was that we might

# ORCHESTRAL MUSIC IN CHURCH.

take an active part in the introduction of orchestral music into Church festival services. This was a subject in which he took great interest, and we played together in a Gregorian Festival in St. Paul's Cathedral, at Harvest Festival Services at St. Stephen's, Lewisham, St. Philip's, Battersea, and the pretty little parish church of Darenth, Kent. Just at the time, however, at which, under the tuition of the late A. J. Phasey, he was becoming master of his instrument, he received the injury to his right hand already described. And though he several times subsequently spoke of resuming his practice, and did occasionally attempt to do so, he never really took up the euphonium again.

Of instrumental music, however, whether himself performing or not, he was always very fond, and, when we lived at Norwood, used regularly to attend the daily performances of the famous orchestral band at the Crystal Palace, under the leadership of Mr. August Manns. With his usual happy knack, too, of succeeding where anyone else would certainly have failed, he managed to obtain an order admitting him to the private rehearsals for the Saturday Concerts; for on Saturday afternoons he was generally engaged, and so could not attend the concerts themselves. But his strong point, as far as music was concerned, lay in the organisation and conducting of choral music. Besides undertaking the precentorship of the seven great festivals which have already been described, he trained and managed the parish choir of Erith, Kent, during a period of several years, and also conducted an amateur

u 2

choral society, which met at private houses, but
occasionally gave public concerts.

His method of conducting was peculiarly his own.
For an occasional false note he cared very little ; but
false time, or false accent, used to trouble him
exceedingly. I have often known him to come downstairs from the top of the house in order to set right a
bar that was not quite correctly played in one of these
respects. In the same way, when practising his choir,
he would have a single line, or even a single bar, of a
hymn repeated again and again until it quite suited his
fastidious requirements. But in this way he got his
choirs to sing with a precision and "swing" seldom
attained by those not under professional tuition, and his
own choir of Erith was famous for miles round for the
high state of excellence to which it attained under his
rule.

Of course there were occasional objections, on the
part of some of the choristers, to a system in which so
much attention was paid to detail; and now and then
very amusing little scenes would result. Of one in particular my father was never tired of telling. Certain
malcontents were greatly dissatisfied with the manner
in which a particular piece of music was taken. Each
had suggestions to make, and of course these different
suggestions were altogether contradictory. At last one
of the members rose and said that, as there was clearly
much dissatisfaction, he had a proposal to make which
he thought would solve every difficulty ; and that was,
that the members of the choir should conduct and

the Rev. J. G. Wood should sing! This proposition
naturally had the effect of closing the controversy; and
ever afterwards the conductor's task was an easier and
a more agreeable one.

Among other amusements my father was very fond
of whist. But it had to *be* whist, true and serious, for
nothing of the "bumble-puppy" order would he endure.
The game was a game to be played in silence, with
due regard for the solemnity of the occupation; failing to return a partner's lead was a great and grievous
fault, requiring to be expiated by due penitence and
contrition, with much self-abasement and promise of
amendment; and a call for trumps unobserved was
regarded with a sort of wistful sorrow, as of one who
feels deeply, but is too saddened to scold. Scold, however, my father never did, at whist or at anything else.
He erred, indeed, if he erred at all, rather on the side of
over-indulgence; and I do not think that in all my life
I more than once heard angry words from his lips—
and then only under circumstances of the very greatest
provocation.

Chess he used to play at one time, and he played it
well; but he gave it up on account of the strain which
it caused on an already sufficiently taxed brain. But of
backgammon—which he played after a fashion of his
own—he was extremely fond; and while he was laid
up with the accident to his hand we played almost every
evening, he having instructed me in the game for
this special purpose. And at one time he devoted
some little attention to mistram, which he tried—with

some success—to raise to the rank of a game of skill
from that of a game of chance. Other indoor games
than these I do not think that I ever knew him to
play.

His principal indoor recreation always consisted in
reading; and in this he was omnivorous. Newspapers,
novels, pamphlets, magazines; grave or serious, witty
or wise; all were seized upon and read in all sorts of odd
moments. And he mostly had his pockets full of odds
and ends of literature which he had picked up at railway
bookstalls. He also, I am sorry to say, had a bad way
of borrowing books, and either forgetting to return them,
or else leaving them somewhere without the slightest
recollection of having done so. In just the same way
he used to leave his linen behind him at the various
places at which he stayed during his lecturing tours;
and some used to come home by parcel post; and some
never came at all. So that after about three tours of
average length he always required a new outfit. With
other people's books, of course, the matter was far more
serious; and I am afraid that his inveterate forgetful-
ness in this respect was the cause of many losses to
many people.

In reading and in conversation alike, there was one
subject with which my father would have nothing whatever to do; and that, strange to say, was the subject of
politics. No matter how fiercely the storm of controversy might wage; no matter how great might be the
excitement throughout the country on some burning
question of the day; no one could elicit his views on

the subject, no one could ever ascertain whether he had any views at all. What the reason for this curious reticence might be I never could discover; but it was very deeply rooted. He impartially read the newspapers of both sides, but invariably kept his opinions to himself. When voting-day came round he went and voted, as became a man and a ratepayer; but no one ever knew to which side his vote had gone. And almost the only way to offend him was to persist in talking to him upon the one subject which he abhorred, or to press for a statement of his views upon matters which all the rest of the world were eagerly and excitedly discussing.

Although we saw but little of him when at home, and although, even upon the few occasions on which he left his study, he was commonly engrossed in the particular subject upon which he happened to be working, my father was always a delightful companion, with the power of thoroughly entering into and sympathising with the hobbies and pursuits of those much younger than himself. In society, too, he was always very popular. He might talk, or he might not: that, to a great extent, was a matter of chance. But, when he once began to enter into conversation, and to interest himself in the topic of discussion, he was one of the pleasantest of talkers, always ready and apt with quotation, anecdote, and illustration, full of humour, and at the same time very careful and accurate in his statements. In the words of one who knew him well, from the time of his matriculation to the end of his life, " he was an excellent conversationalist, and no matter how

many hours he had been toiling, he always seemed fresh
for his work, and pleased to have a friendly chat."

Perhaps the best tribute to his popularity in society
was the fact that he was always equally welcomed by
all, of both sexes and all ages. Everywhere that he
went he made new and staunch friends; and the number
and tone of the sympathetic letters received after his
death bore eloquent testimony to the universal favour
with which he was regarded, and the deep and real sense
of the loss which all who knew him felt that they had
sustained.

# INDEX.

                THE  END.


PRINTED BY CASSELL & COMPANY, LIMITED, LA BELLE SAUVAGE, LONDON, E.C.